THE TEACHING OF SCIENCE:

21st
CENTURY
PERSPECTIVES

THE TEACHING OF SCIENCE:

21st CENTURY PERSPECTIVES

RODGER W. BYBEE

National Science Teachers Association

Arlington, Virginia

National Science Teachers Association

Claire Reinburg, Director
Jennifer Horak, Managing Editor
Andrew Cooke, Senior Editor
Judy Cusick, Senior Editor
Wendy Rubin, Associate Editor
Amy America, Book Acquisitions Coordinator

ART AND DESIGN
Will Thomas Jr., Director
Joe Butera, Senior Graphic Designer, cover and interior design
Cover photo by Igor Gorelchenkov

PRINTING AND PRODUCTION
Catherine Lorrain, Director
Jack Parker, Electronic Prepress Technician

NATIONAL SCIENCE TEACHERS ASSOCIATION
Francis Q. Eberle, PhD, Executive Director
David Beacom, Publisher

Library of Congress Cataloging-in-Publication Data
Bybee, Rodger W.
  The teaching of science : 21st-century perspectives / by Rodger W. Bybee.
      p. cm.
  Includes bibliographical references and index.
  ISBN 978-1-936137-05-3
  1. Science--Study and teaching. 2. Science teachers--Training of. 3. Science teachers--Vocational guidance. I.
National Science Teachers Association. II. Title.
  LB1585.B94 2010
  507.1--dc22
                            2010020234

eISBN 978-1-936137-61-9

Contents

For
Sumner Clark Eakins
and his young friends in the class of 2024

They represent the generation that will benefit from teaching science with
21st-century perspectives.

Preface

Science teachers at all levels—elementary, middle, and high school—confront diverse issues and requirements, all of which can divert teachers' time and attention from the fundamental task of helping students learn science. In their need to focus on the immediate tasks, teachers ask for lessons that will get them through the day or week. Although they certainly sense the need, if not the obligation, to pause and ask essential questions, they seldom have the time for reflection: What science content and processes are important for students to learn? How can I organize experiences to facilitate student learning? How will I know what students have learned? What knowledge and skills do I have to have to help students learn? To be clear, these questions may have variations, but they center on the instructional core that all teachers recognize as fundamental to their work as professionals.

The content and themes of the chapters in this book may be used to reflect on issues basic to the teaching of science. The topics and discussions in the book lend themselves to "summer reading" or professional development discussions with colleagues. This book neither emphasizes nor presents activities for teaching. The themes mostly address why to teach science and what is important to teach. Answering the *why* and *what* questions contributes to constructive responses to the *how* questions. Where possible and appropriate, I have provided references and resources that will help science teachers with their daily, weekly, and yearly tasks as professionals.

I have been honored to present several major lectures at National Science Teachers Association (NSTA) meetings. The original titles as well as the lecture locations and dates are listed on the following page. My practice is to prepare a written essay for the lectures. In all but two cases, those essays have not been published. Upon rereading the lectures, I realized two things. First, I tried to present ideas about curriculum and instruction in a style appropriate for science teachers. Second, the lectures made connections between the past and future. The latter occurred because the lectures were named for individuals—Robert H.

Karplus, Paul F-Brandwein, and Robert H. Carleton—who have made significant contributions to science education and influenced my career.

In preparing the chapters for this book, I have maintained the themes set forth in the original lectures. Because the lectures were presented several years apart, I took the liberty to change the sequence and update the chapters by adding contemporary information, eliminating some redundancies, and adding resources and references. In addition, I moved to the prologue the personal introductions about the persons for whom the lectures were named.

The first chapter introduces the subsequent chapters with major themes and an emphasis for the book. I also set forth the themes of curriculum and instruction as they relate to science teachers.

NSTA Lectures
Original Titles, Locations, and Dates

Reflections on Curriculum and Instruction
The Robert H. Karplus Lecture
NSTA National Conference
San Diego, California
March 29, 2002

The Teaching of Science:
Content, Coherence, and Congruence
The Paul F-Brandwein Lecture
NSTA National Conference
Philadelphia, Pennsylvania
March 29, 2003

Teaching Science and
Fulfilling National Aspirations:
The Critical Role of Curriculum Reform
Life Members Lecture
NSTA National Conference
St. Louis, Missouri
March 30, 2007

The Robert H. Carleton Lecture
NSTA National Conference
Boston, Massachusetts
March 28, 2008

Scientific Literacy and Environmental Issues:
Insights from PISA 2006
The Paul F-Brandwein Lecture
NSTA National Conference
Boston, Massachusetts
March 29, 2008

The second chapter is based on my 2003 Paul F-Brandwein Lecture. In this chapter, I introduce Brandwein's original themes of Substance, Structure, and Style and connect these themes to contemporary Content, Curricular Coherence, and Congruence. The bases for these themes are national standards, research on learning, and the role of inquiry in science instruction.

To the directors of the Paul F-Brandwein Institute, my colleagues, and especially my friend for more than 30 years, John Padalino, I extend my appreciation for the opportunity to present the 2003 Paul F-Brandwein Lecture. I took it as a great honor to present a lecture in memory of Paul F-Brandwein—a great environmentalist and a great science teacher. On several occasions, I had the opportunity to talk with Paul F-Brandwein and always found him to be personable and understanding of a young professional who wanted to understand his views on science education, environmental education, and gifted students who had interests in scientific careers.

If I may add a personal note of acknowledgment, I have known and worked with John (Jack) Padalino since our days in graduate school at New York University. He, like Paul, is a great environmentalist and distinguished educator. For years he worked to see that inner-city students participated in environmental education experiences that many would not have had without his extraordinary efforts. Jack has constantly reminded me that science education is largely political and mostly local. This bit of wisdom has been helpful on numerous occasions, as my professional work has encompassed local, national, and international perspectives.

The third chapter is based on my 2002 Robert H. Karplus Lecture. The essay includes an introduction to the influence Karplus had on curriculum development and ideas that we used at BSCS. This chapter also includes a detailed discussion of the BSCS 5E Instructional Model and its origins from the Karplus learning cycle.

I truly appreciated the opportunity to reflect on curriculum and instruction in general and the contributions of Robert H. Karplus in particular. I was deeply honored, as 2002 marked 50 years since the initial work on the Science Curriculum Improvement Study (SCIS). I also was thankful for a chance to discuss a bit of the history of science education.

Although I did not realize it at the time, I began reflecting on curriculum and instruction in 1968 when I spent a memorable week visiting SCIS. This was the first time I met Bob Karplus. During the next 13 years, I had numerous opportunities to visit with Bob, attend his presentations, read his publications, and use materials that he and his colleagues developed, including the SCIS materials. It would be inappropriate to leave the impression that we had a deep and enduring friendship, but Robert Karplus did have a profound and lasting influence on my career as a science teacher, curriculum developer, and educator. His influence came less through personal interaction and more through his intellectual endeavors, specifically his reflections on curriculum and instruction.

The 2008 Robert H. Carleton Lecture provides the content for Chapter 4. The chapter centers on the themes of teaching science as inquiry. After a brief introduction to the history of inquiry in science education, I use the national standards as the basis for a detailed discussion of inquiry as learning outcomes and teaching strategies. The concluding sections discuss the role of inquiry and preparation of 21st-century skills.

I delivered a second Paul F-Brandwein Lecture in 2008. This lecture is the basis for the fifth and sixth chapters. In that lecture, I used the theme of scientific literacy to introduce the Program for International Student Assessment (PISA), in which science was emphasized in 2006. The specific discussion centers on environmental themes that were assessed in the 2006 PISA.

The opportunity to present the 2008 Paul F-Brandwein lecture left me with no small humility and great honor. I acknowledge all directors of the Paul F-Brandwein Institute, especially those I have known and worked with and admired for years: Keith Wheeler, Alan Sandler, Cheryl Charles, Marilyn DeWall, and William Hammond.

Chapters 7 and 8 are based on my 2007 Life Members Lecture, in which I address 21st-century issues and link ideas from PISA 2006 science to the contemporary need for curriculum reform. Presenting the Life Members Lecture had significant personal meaning for me because it represented my 40th anniversary as a member of NSTA. I used the occasion to talk about two themes that have been central in my career as a member of NSTA: The first theme centers on fulfilling national aspirations, and the second theme addresses the critical role of instructional materials and curriculum reform.

In the epilogue, I address the need for leadership and the responsibilities for continued reform in science education.

Like any author, I must acknowledge the fact that many individuals contributed to the themes and ideas expressed in this book. I have benefitted greatly from my recent work on the PISA and discussions with members of the Science Expert Group, and especially my colleague Barry McCrae from the Australian Council for Educational Research (ACER). Barry continually asked for clarification and a rationale for ideas that became part of the PISA 2000 science assessment. Many of those ideas are integral to the themes in this book.

These NSTA lectures were presented during my tenure as executive director of BSCS. Support and encouragement from Pam Van Scotter, Nancy Landes, Joe Taylor, and Janet Carlson were not only helpful but also vital, and I acknowledge their assistance.

Several colleagues are part of a special NSTA meeting. Discussions during these yearly meetings have broadened and deepened my understanding of science education. Here I acknowledge Mark St. John, Harold Pratt, and

David Heil for their understanding of the personal and professional lives of science educators.

There is a special note of appreciation for Kathryn Bess, who listened, questioned, and clarified ideas that became central to the lectures and this book. This book's emphasis on science teachers and teaching is due largely to Kathryn's wise counsel.

I thank Claire Reinburg of NSTA for her support from the beginning proposal to the final product and Wendy Rubin for her contributions and seeing this manuscript through final production.

Finally, my assistant, Byllee Simon, contributed in numerous ways to the final manuscript for NSTA. I continue to appreciate her interest in, contributions to, and support of my work.

During my career I have been thankful for numerous interactions with science teachers. Their wisdom and experience have both tempered some ideas and embellished others. I certainly thank them and extend my appreciation for their ideas. They are the central hope for helping students realize their future as citizens, some of whom will be scientists and engineers.

Rodger W. Bybee
Golden, Colorado
February 2010

Prologue
Connecting the Past and Future

In the preface, I mentioned the fact that I knew the individuals for whom the NSTA lectures were named—Paul F-Brandwein, Robert Carleton, and Robert Karplus—and who had a great influence on my career. As work on this book continued, I thought it important to provide readers with a brief introduction to these individuals. The following discussion and this book connect these 20th-century leaders to future generations of science teachers as they themselves become the 21st-century leaders.

Paul F-Brandwein: Scientist, Environmentalist, and Curriculum Designer

The Brandwein Lectures both acknowledge Paul F-Brandwein's long and distinguished career, including serving on the Steering Committee of the Biological Sciences Curriculum Study (BSCS) from the late 1950s into the 1960s. Paul F-Brandwein directed the Gifted Student Committee at BSCS and was responsible for initiating a program on student research problems. He felt deeply about giving students the opportunity to engage in scientific inquiry as a means to encourage their future careers as scientists.

Paul F-Brandwein played a key role in BSCS's early publications for gifted students. He was a member of the BSCS Steering Committee and the Gifted Committee from 1959 to 1962 and a member of the Special Student Committee from 1962 to 1963. I also would note that Harcourt Brace, the company for which Paul was a senior editor and an education consultant, published BSCS's *Biological Sciences: An Inquiry Into Life*, known as the BSCS "Yellow Version."

Brandwein had impressive credentials in addition to his position at Harcourt: consulting science editor to Science Research Associates; associate director of the Joint Council on Economic Education with special responsibility as director of its Conservation and Resource-Use Project; associate editor of NSTA's journal *The*

Science Teacher; and president of the Federation of Science Teachers of New York. He taught in New York City high schools for 15 years and was chairman of a science department for 10 of those years. Brandwein also had 15 years of college teaching experience, including positions at New York University, Teachers College, Columbia University, and Harvard University.

Among his publications before his work for BSCS were *The Gifted Student as Future Scientists*; *You and Science*; *The Physical World*; *Teaching High School Science: A Book of Methods*; *Teaching High School Science: A Sourcebook for the Biological Sciences*; and *Teaching High School Science: A Sourcebook for the Physical Sciences*.

A Biology Education for Gifted Students

Brandwein was especially perceptive in his observations about the gifted student, noting at a Steering Committee meeting that identifying the gifted student was one of the most important problems for science teachers. He said that we frequently confuse "brightness" with "giftedness." A bright student accepts what is presented by the instructor; the gifted student may question what is given to him by the teacher and may not fit into the classroom emotionally or otherwise. Dr. Anne Roe of the Graduate School of Education at Harvard University was a member of the BSCS committee and a colleague of Brandwein. She studied the intellectual and emotional characteristics of gifted students and found that most of them are dissatisfied with the present explanation of reality and continually search for more satisfying explanations (Grobman 1969). His concern with providing challenging science experiences for gifted students led Brandwein to propose a program of BSCS materials.

The Gifted Student Committee agreed to organize materials that could be used by high school science teachers to encourage the work of highly talented students, especially in biology. The plans called for assembling about 300 investigations that these students might conduct. The investigations were conceived as original research problems for which solutions were not yet available in the literature and were intended to take several years of work to accomplish. After the students completed their research investigations, they would write up their results and submit them to BSCS for editing; the results would then be returned to the student for approval and finally forwarded to an appropriate journal for publication under the student's name. The Gifted Student Committee planned to enlist the collaboration of biologists throughout the country in preparing brief outlines of research projects for these students (BSCS 1960).

During the 1960 Summer Writing Conference in Boulder, Colorado, six members of the Gifted Student Committee worked on the new materials. Members of that committee included Paul Brandwein; Hurbert Goodrich, Wesleyan University; Jerome Metzner, Bronx High School of Science; Richard Lewontin, University of Rochester; Evelyn Morholt, Fort Hamilton High School, Brooklyn, New York; and Walter Rosen, Marquette University.

Research Problems for Biology Students

The Gifted Student Committee selected and edited 100 proposed research problems from research biologists, and these were eventually published in a volume titled *Biological Investigations for Secondary School Students*. The book included a preface that oriented gifted students to the selection and use of a prospectus and a bibliography of general and specific references. The committee also planned to develop a means of evaluating the use of these proposed problems by participating schools.

In anticipation of teaching science as inquiry, a theme developed in this book, I quote from the introduction to *Biological Investigations for Secondary School Students*:

> *These one hundred ideas for investigation were developed to bring you the opportunity to gain experience in the art of investigation. You probably will not find "answers" to the problems they pose in textbooks, nor do we expect you will find a possible avenue to their solution in the references appended to each one. However, the careful thought and zealous work, the imaginativeness and inventiveness you will bring to the investigation, will yield you two or three years of exciting work. You may even be fortunate enough to discover a new fact, a new relationship, a new technique; you may be the first to know something no one before you has known. You may experience the thrill which comes to the scientist, the thrill of discovery, and more than that, you may have the joy of sharing your discovery with others. (BSCS 1961)*

In 1962, the activities of the BSCS Committee on the Gifted Student involved changing its name to the Committee on the Special Student to include students at both ends of the ability range. A subcommittee chaired by Evelyn Klinckmann of San Francisco College for Women defined unsuccessful learners to include the 20 to 30% of students taking high school biology who had difficulty with BSCS biology. At the 1963 Summer Writing Conference, the committee proposed producing materials for those students who had not been successful in field tests of BSCS programs.

By 1964, under Brandwein's leadership, the Committee on the Special Student had written three publications, including *Teaching High School Biology: A Guide to Working With Potential Biologists* (Brandwein et al. 1962). This volume was developed for teachers working with strong biology students. It contained material on the characteristics of the gifted student (with particular reference to science); strategies for encouraging the development of an art of investigation; promising practices in the teaching of students of high ability in biology as observed in U.S. classrooms; and an introduction to the use of the library as well as a bibliography on "giftedness." Additionally, two volumes of research problems in biology were prepared. Each of these paperback volumes had 40 investigations that were useful for originating problems for research on the school level (Grobman 1969).

A Conceptual Framework for BSCS

Paul F-Brandwein had significant influence on the conceptual framework used at BSCS. In a 1976 article titled "Reflections on the Early Days of BSCS," Bentley Glass had this to say after an introduction about organisms and the levels of organization used in the design of BSCS programs:

> Especially, we agreed to select and emphasize a limited number of great biological concepts, or themes, that would run clearly throughout every phase of the treatment in every version, or program. The nine themes we chose, a procedure in which Paul Brandwein played a leading part, are so well known it is unnecessary to itemize them, except in the form of the diagram which provides our matrix of organizing ideas. (Glass 1976, pp. 3–4)

You can see in this quotation the importance that Brandwein placed on major conceptual ideas, in this case for the discipline of biology. I thought this quotation especially appropriate because it shows Paul's leadership at BSCS and provides connections to other sections of this book. His ideas influenced the other founders and early development of BSCS. Indeed, his influence continues to this day and will do so into the future.

As a gifted teacher himself, Brandwein clearly had a major influence on BSCS programs for the exceptionally talented science student. He came to BSCS well aware of the limitations of the lecture and of existing textbooks and was determined to help transform science education. To quote Calvin Stillman,

> The role of the warm mentor is fundamental in Paul's work. The younger person has to identify himself, and once he does so, the mentor is the strong person who helps the young one to find out [through original work] what it means to be a scientist. For Paul, science was the system of constructing a hypothesis and testing it carefully, with no sense of failure if the hypothesis turns out to be wrong. (Stillman 1997)

There was a second aspect of Paul's career, conservation. His activity as a conservationist was lifelong; indeed, it has extended beyond his life in the form of property he and his wife, Mary, bequeathed (as the Rutgers Creek Wildlife Conservancy) to an organization committed to students, teachers, and scientists interested in the environment and natural systems. That conservancy has been administered through an affiliation with the Pocono Environmental Education Center at Dingman's Ferry, Pennsylvania. John Padalino directed the center until his retirement.

Robert H. Carleton: Science Educator, Administrator, and Education Leader

In the late 1960s, as a graduate student at the University of Northern Colorado, I met and had several opportunities to visit with Robert Carleton. He quietly listened to my questions, which I am sure were simple if not naïve, and talked about the role of the National Science Teachers Association (NSTA) in local, state, national, and international science education. Only later did I realize the depth and breadth of his leadership.

During his undergraduate and graduate studies in science education at two major universities, Carleton was elected to Phi Beta Kappa. For more than four decades, Robert Carleton contributed to science education as a high school teacher, university professor, and executive secretary of NSTA. He served as executive secretary of NSTA for 25 years. During his tenure as executive secretary of NSTA—one of the foremost leadership positions in the field of science teaching—Carleton demonstrated the unique abilities of creative and sound ideas combined with the energy and political wisdom to carry those ideas to fruition. Working harmoniously with diverse elected officers of NSTA, he was a model of national leadership.

In his years as NSTA's executive secretary (1948–1973), Robert Carleton participated in numerous national and international committees, conferences, and advisory groups concerned with supporting science teachers and advancing science education. He also was the author of more than a dozen textbooks in science, part of his many contributions to the teaching field during his career.

Robert H. Karplus: A Science Teacher and Education Leader

In the late 1950s and early 1960s, a number of scientists became actively involved in science education in general and curriculum development in particular. Some of the names may be familiar: Jerrold Zacharius, Glenn Seaborg, David Hawkins, Bentley Glass, Arnold Grobman, and John Moore. Robert Karplus joined the science education community when he became interested in elementary school science. This was in fact Bob's second career. His first career was in theoretical physics and included work at the Institute for Advanced Study in Princeton, New Jersey; Harvard; and the University of California, Berkeley. As a theoretical physicist, Karplus had a brilliant and exceptional career, which he left to take on the challenges of curricular reform in science education (Fuller 2002).

As a father of seven children, Bob's responsibility as a parent combined with his curiosity and interest in science naturally extended to schools. In 1958, Bob visited his daughter Beverly's second-grade classroom to teach several science lessons. Bob gave a physics lecture to second graders. You can only imagine the children's response. Karplus took this encounter seriously, as he wanted children to understand the wonders of science and appreciate the excitement of discovery that he had experienced as a scientist.

> *I cannot resist telling two other stories about Karplus—the learner as teacher. Robert Karplus placed the toy truck in front of a child. He rolled the truck slowly across the desk. "Did the truck move?" he asked. "No," replied the child.*
>
> *(It is difficult to learn the fundamental concepts of motion when an object that goes from one location to another does not move. Perhaps he had misunderstood. He moved the truck back to its starting position. Again, he slowly rolled the toy truck across the desk to a new location.)*
>
> *"Did the truck move?" he asked again. "No," the child replied once again. "Can you explain to me why you say the truck did not move?" Karplus asked. "It did not move," responded the child triumphantly. "You moved it!"* (Fuller 2002, p. 301)

Another classroom experience always touches the heart and brings a smile to any science teacher. Karplus believed it was important to see phenomena and interpret investigations from a scientific point of view. Karplus designed a series of activities to help children understand that many processes of change in a system eventually come to a balance point when the system reaches equilibrium. At the conclusion of his investigations, one boy announced to Professor Karplus, "I know something that will go on forever. You will keep on talking forever" (Karplus and Thier 1967). I can only imagine Bob, with that great smile and a twinkle in his eye, changed to a new topic.

Jerome Bruner paid a great tribute to Robert Karplus, the science teacher, when he had this to say about Bob:

> *His ideas about how to teach science were not only elegant but from the heart. He knew what it felt like "not to know," what it was like to be a "beginner." As a matter of temperament and principle, he knew that not knowing was the chronic condition not only of a student but of a real scientist. That is what made him a true teacher, a truly courteous teacher. What he knew was that science is not something that exists out there in nature, but that it is a tool in the mind of the knower-teacher and student alike. Getting to know something is an adventure in how to account for a great many things that you encounter in as simple and elegant a way as possible.* (Fuller 2002, p. 321)

During this period of initial work in science education (generally 1958–1963), Karplus worked with other University of California, Berkeley, faculty on the Elementary School Science Project (ESSP) and visited the Elementary Science Study (ESS). He also participated in a summer curriculum development for MINNEMAST, a mathematics and science program at the University of Minnesota, Minneapolis.

In the course of these experiences as a teacher and curriculum developer, Karplus pondered several insightful questions. First, how can one create learning experiences that achieve a connection between the pupil's intuitive attitudes and the concepts of the modern scientific point of view? Second, how can one determine what the children have learned? Third, how can one communicate with the teacher so that the teacher can in turn communicate with the pupils (Karplus and Thier 1967, p. 11)? Such questions led Karplus to a personal study of psychology, in particular, the work of Jean Piaget. Embedded in these questions are ideas that anticipate the contemporary science of learning and curriculum development by extension.

By 1963, Robert Karplus had the professional experience with science, students, and curriculum study; the personal time to reflect on fundamental questions about curriculum and instruction; and the opportunity to develop his ideas in the Science Curriculum Improvement Study (SCIS).

I truly appreciated the opportunities these lectures provided to reflect on the leadership and contributions of Paul F-Brandwein, Robert Carleton, and Robert Karplus. The ideas they shared about science concepts and processes, curriculum, instruction, assessment, professional development, and management of projects and organizations were formative at the time and continued to develop as I grew as a professional. The chapters in this book both honor their legacies and connect their ideas formed in the 20th century to 21st-century perspectives.

The Teaching of Science: Contemporary Challenges

In the early decades of the 21st century, science teachers face some problems unique to our times and some common to all eras. Among the challenges that require attention and leadership by science teachers are

- achieving scientific literacy,
- reforming science programs,
- teaching science as inquiry,
- improving science teachers' knowledge and skills, and
- attaining higher levels of achievement for all students.

These challenges present a variety of issues that require leadership at all levels within the science education community, but especially by science teachers. Some of the issues will be with us for the relatively brief time of political administrations, and some challenges have a longer and deeper educational standing. Although the challenges extend to all components of science education, my emphasis in this book will be on the core of instructional practice. This, it seems to me, is essential and centers on the science teacher.

The challenges center on longstanding themes in science education— scientific literacy, science programs, science as inquiry, professional development, and student achievement. These themes unite diverse topics and discussions in the chapters that follow. This chapter continues with an introduction to the core of education practice.

The Core of Instructional Practice

In the coming decade, and in future decades for that matter, leaders must direct attention to the core elements of science education. You may ask what I mean by the core elements of science education. As an initial definition, I suggest Richard Elmore's:

By "the core of educational practice," I mean how teachers understand the nature of knowledge and their students' role in learning, and how these ideas about knowledge and learning are manifest in teaching and class work. The "core" also includes structural arrangements of schools, such as the physical layouts of classrooms, student grouping practices, teachers' responsibilities for groups of students, and relations among teachers in their work with students, as well as processes for assessing student learning and communicating it to students, teachers, parents, administrators, and other interested parties. (Elmore 2004, p. 8)

Several features of this quotation are critical for science teachers. First, Elmore cites the importance of teachers' understanding the nature of scientific knowledge and their students' roles in learning science. This feature centers on the science teacher and has direct implications for professional development. Second, he underscores how ideas about scientific knowledge and students' learning of science are realized in the classroom. I translate this to the traditional categories of curriculum and instruction. Third, he recognizes broader programmatic and systemic factors such as classroom student grouping, teachers' responsibilities and collegiality, and finally the process of assessment of student learning. Certainly assessment will be on the teachers' agenda for the foreseeable future. The basic categories of the education core can be identified using the traditional terms of *curriculum*, *instruction*, and *assessment*, with the underlying foundations of student learning and the continuous professional development of science teachers.

In 2009, Elmore again addressed this challenge in "Improving the Instructional Core" (2009). Although Elmore uses the word *instructional* (instead of *educational*), this clear statement makes a fundamental point for this discussion.

There are only three ways to improve student learning at scale: You can raise the level of content that students are taught. You can increase the skill and knowledge that teachers bring to the teaching of that content. And you can increase the level of students' active learning of the content. (Elmore 2009, p. 249)

You can see that focusing on the instructional core recognizes the complex and difficult work of science teaching and student learning. Put simply, the role of science teaching is too important to avoid and too critical to misrepresent. Figure 1.1 presents the instructional core.

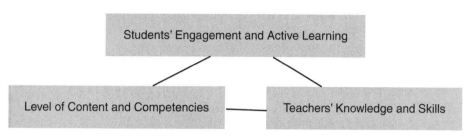

Figure 1.1

Instructional Core of Practice for Science Teaching

Students' Engagement and Active Learning

Level of Content and Competencies — Teachers' Knowledge and Skills

Unfortunately, despite the unmistakable importance of centering on the instructional core and the associated challenges, there often are misdirected emphasis and undue attention on single initiatives that are largely political and distinctly different from what I am describing as the instructional core. Most worrying is the contemporary emphasis on assessment, which does not account for the insights and possibilities that science teachers can bring to student learning.

In crafting responses to the challenges, I argue that science teachers and others in the larger science education community need a serious, informed, and sustained discussion of what it will take to address contemporary challenges: scientific literacy, reforming science programs, teaching science as inquiry, providing professional development, and attaining higher levels of student achievement. The remainder of this chapter and the rest of this book offer a foundation for that discussion.

Asking the Right Questions

Any strategy to improve science teaching and student learning must respond to the aforementioned challenges and reform all components of the instructional core. This reform begins by asking the right questions, clarifying the issues, and reinforcing the strengths of science teachers. Broadly speaking, the critical questions associated with the challenges are the following:

- *Achieving scientific literacy: Tools for future scientists or citizens?* Should school science programs emphasize an education that will set a foundation for scientific study, or one that will help future citizens with science-related life situations? What is the balance of content and experiences that will help students achieve knowledge of science, and about science, as a human enterprise?
- *Reforming science programs: Current science programs or research-based approaches?* How can science teachers unite advances on knowledge about teaching and learning with curriculum and instruction? What will it take

to bridge theory and practice in science education? What are the responsibilities of researchers and teachers to see that new knowledge is applied in the classroom?

- *Teaching science as inquiry: Teach science content or process?* Is it necessary to separate science content and process? What should students know and be able to do relative to scientific procedures? Can science classrooms incorporate both? If not, how can both goals be attained?

- *Improving science teachers' knowledge and skills—professional development: Teach scientific knowledge or pedagogical skills?* What will be most helpful for science teachers? How can we productively move beyond lessons "to do on Monday"? What do teachers really need to do their job?

- *Attaining higher levels of student achievement: Assessment for individual development or international comparison?* Can we respond to both the needs of teachers and policy makers? Can we reduce or close the achievement gap?

Scientific Literacy: Achieving Goals for All Citizens

The term *scientific literacy* expresses the major goal of science education—advancing individual development and satisfying society's aspirations through science education. I think it is reasonable to say that most science teachers would support a goal such as "scientific literacy for all learners." Such a phrase embodies the highest and most admirable goals of science education. Let's examine the idea of scientific literacy.

A Brief Introduction to Scientific Literacy

Use of the term *scientific literacy* most likely began with James Bryant Conant in the 1940s (Holton 1998) and was elaborated on for educators in a 1958 article by Paul DeHart Hurd titled "Science Literacy: Its Meaning for American Schools." Hurd described the purpose of scientific literacy as an understanding of science and its applications to social experience. Science had such a prominent role in society that economic, political, and personal decisions could not be made without some consideration of the science and technology involved (Hurd 1958).

Achieving Scientific Literacy

In the 50 years since Hurd's article, the term *scientific literacy* has been used extensively to describe the purposes, policies, programs, and practices of science education. The term *scientific literacy*, however, is not the reality of science education. Academic researchers debate the real meaning of the term, classroom teachers claim their students are attaining scientific literacy, and national and international assessments provide evidence that somewhere between the abstract purposes of science education and concrete practice in science classrooms, the education community is not achieving the goal, at least in the United States.

In Hurd's article, he linked scientific literacy to social experience and provided a rationale in economic, political, and personal contexts. Hurd made a clear connection between science and citizenship. In contrast, scientific literacy is not exclusively preparation for a professional career, although Hurd's connection does not preclude scientific and technological careers. Scientific literacy—as it is manifest in education policies, programs, and practices—has the explicit goal of preparing students for life and work as citizens.

In my 1997 book *Achieving Scientific Literacy: From Purposes to Practices*, I attempted to clarify what is meant by *scientific literacy* and the use of the term as a slogan and metaphor. In that discussion, I proposed that scientific literacy clarified the general purposes of science education; implied the same standards for all students; illustrated different emphases for curriculum, instruction, and assessment; represented a continuum of understandings and abilities; incorporated multiple dimensions; and included both science and technology.

At the time, the term *scientific literacy* often was used as the basis for judgments about individuals or society. For example, one commonly heard or read about individuals or groups being labeled scientifically illiterate—based on the observation that an individual did not know the difference between, for example, an atom and a molecule, a mineral and a rock, or an organism and a species. Because of this observation, I proposed a model that assumed scientific literacy was continuously distributed in a population and had multiple dimensions. At the extremes, there were small numbers of individuals who were scientifically literate and illiterate. But within the greater population there was a distribution of individuals who demonstrated varying degrees of scientific literacy. Variation in the population was a function of factors such as age, cognitive development, school curricula, and life experiences. Defining characteristics of the scientific literacy continuum included illiteracy, nominal literacy, functional literacy, conceptual and procedural literacy, and multidimensional literacy (Bybee 1997).

In 1997, Thomas Koballa, Andrew Kemp, and Robert Evans published an article in which they also presented a spectrum of scientific literacy (e.g., illiteracy to highest levels of understanding), multiple domains (e.g., biology, history of science), and personal attitudes attached to pursuing scientific literacy (e.g., low to high). The spectrum of scientific literacy described by Koballa, Kemp, and Evans included, clarified, and elaborated on many aspects of the ideas I described.

Contemporary Perspectives

This discussion presents a brief review of the literature on scientific literacy, yet several other authors and reports should be mentioned. George DeBoer (2000) has provided an excellent historical and contemporary review of scientific literacy. In 2006, Robin Millar addressed historic and defining issues of the term before outlining the role of scientific literacy in a program called Twenty First Century Science course. Now that course stands as a model curriculum. In discussing the distinctive features of the curriculum, Millar points out the novel content (e.g., epidemiology, health), broad qualitative understanding (e.g., whole explanations), and a strong emphasis on ideas about science (e.g., understanding the nature of science).

Two essays stand out when discussions turn to contemporary science education and the challenges of attaining higher levels of scientific literacy. In his essay "Science Education for the Twenty First Century," Jonathan Osborne (2007) makes a clear case that regardless of the use of scientific literacy as a stated aim, contemporary science education is primarily "foundationalist" in that it emphasizes educating future scientists versus educating future citizens. Osborne is critical of the domineering emphasis on the foundationalist orientation and diminished recognition of an orientation toward future citizens. Among other outcomes, Osborne argues that such an emphasis results in students having negative attitudes toward science.

The second noteworthy essay is by Douglas Roberts and was published in *Handbook of Research on Science Education* (Abell and Lederman 2007). Roberts identifies a continuing political and intellectual tension with a long history in science education. The two politically conflicting emphases can be stated in a question: Should curriculum emphasize science subject matter itself, or should it emphasize science in life situations in which science plays a key role? Roberts refers to answers to the former as Vision I and answers to the latter as Vision II. Vision I looks within science, while Vision II uses external contexts that students are likely to encounter as citizens. The ideas presented by Roberts also form the central theme of *Promoting Scientific Literacy: Science Education Research in Transaction*, the proceedings of a symposium held at Uppsala University in Sweden (Linder, Ostman, and Wickman 2007). In the context of this discussion, Vision II is the emphasis that is consistent with early discussions by Paul DeHart Hurd and the contemporary views of Osborne (2007) and Millar (2006). We should be very clear that this emphasis differs significantly from that in the majority of national and state standards and assessments and school science programs.

In PISA 2006, a contemporary international assessment, the essential qualities of scientific literacy include the ability to apply scientific understandings to life situations involving science. The central point of the PISA 2006 science assessment can be summarized as focusing on scientific competencies that clarify what 15-year-old students should know, value, and be able to do within reasonable and appropriate personal, social, and cultural contexts.

The PISA Definition

For purposes of PISA 2006, scientific literacy refers to four interrelated features that involve an individual's

- scientific knowledge and use of that knowledge to identify questions, to acquire new knowledge, to explain scientific phenomenon, to draw evidence-based conclusions about science-related issues;
- understanding of the characteristic features of science as a form of human knowledge and [i]nquiry;
- awareness of how science and technology shape our material, intellectual, and cultural environments; and
- willingness to engage in science-related issues, and with the ideas of science, as a constructive, concerned, and reflective citizen (OECD 2006).

In this new century the science education community must pause and again ask the Sisyphean question: What should the scientifically and technologically literate person know, value, and do—as a citizen? The question is not new; it was the topic of a 1985 yearbook of National Science Teachers Association (Bybee 1985). Identifying what citizens should know, value, and be able to do in situations involving science and technology may seem simple. Answering the Sisyphean question opens the realms of scientific understanding, but it also indicates a qualifier—citizens. As citizens, what *knowledge* is most appropriate? An answer to this question certainly includes basic science concepts, but that knowledge must be applied in contexts that citizens encounter in life. In addition, people often encounter situations that require some understanding of science as a process that produces knowledge and proposes explanations about the natural world.

What is important for citizens to value about science and technology? An answer will include the role and contributions to society of science, and of science-based technology, and their importance in many personal, social, and global contexts. It seems essential that citizens have an interest in science, support the process of scientific research, and act responsibly toward issues that relate to science and technology—for example, health, use of natural resources, and environmental quality.

What is important for individuals to be able to *do* that is science related? People often have to base decisions on evidence and information, evaluate claims made by others on the basis of the evidence, and distinguish personal opinion from scientific (i.e., evidence-based) statements.

In general, citizens do not judge the worth of theories or advances in science. But they do make decisions based on the facts in advertisements, evidence in legal matters, information about their health, and issues concerning local environments and natural resources. An educated person should be able to distinguish questions that can be answered by scientists from problems that can be solved by science-based technologies.

Science Programs: Incorporating Research-Based Approaches Into Curriculum and Instruction

Enhancing student achievement will rely on designing curricula based on research that has advanced our understanding of how students learn science. The following discussion reviews contemporary research that has implications for science teaching and learning. Several examples of curriculum and instruction are provided.

Research on Learning

The National Research Council (NRC) reports *How People Learn: Brain, Mind, Experience, and School* (Bransford, Brown, and Cocking 2000), *How People Learn: Bridging Research and Practice* (Donovan, Bransford, and Pellegrino 1999), *How Students Learn: Science in the Classroom* (Donovan and Bransford 2005), *Taking Science to School: Learning and Teaching Science in Grades K–8* (Duschl, Schweingruber, and Shouse 2007), and *Ready, Set, Science: Putting Research to Work in K–8 Science Classrooms* (Michaels, Shouse, and Schweingruber 2008) present a major synthesis of research on human learning. I will also note *Learning Science and the Science of Learning* (Bybee 2002), a volume I edited for the National Science Teachers Association (NSTA) that presented many of these findings for science teachers. Three findings from the NRC reports, in particular, have both a solid research base and clear implications for science curricula and instruction.

Students Come to Class With Preconceptions

The following findings are from *How People Learn: Bridging Research and Practice* (Donovan, Bransford, and Pellegrino 1999):

> *Students come to the classroom with preconceptions about how the world works. If their initial understanding is not engaged, they may fail to grasp the new concepts and information that are taught, or they may learn them for purposes of a test but revert to their preconceptions outside the classroom.* (p. 20)

The curricular implications of this first finding relate to the structure of experiences that draw on our students' current understandings, bring about some sense of the inadequacy of the ideas, and provide opportunities and time to reconstruct their ideas so they are consistent with basic scientific concepts.

Competence Requires a Conceptual Framework

A second finding refers to the conceptual foundation of a curriculum:

> *To develop competence in an area of inquiry, students must: (a) have a deep foundation of factual knowledge, (b) understand facts and ideas in the context*

of a conceptual framework, and (c) organize knowledge in ways that facilitate retrieval and application. (Donovan, Bransford, and Pellegrino 1999, p. 10)

A science curriculum should incorporate fundamental knowledge and be based on and contribute to students' development of a strong conceptual framework. Research comparing performance of novices and experts, as well as research on learning and transfer, shows that experts draw on a richly structured information base. Although factual information is necessary, it is not sufficient. Essential to expertise are both the mastery of concepts that allows for deep understanding and a framework that organizes facts and information.

Thinking About Problems

Finally, there is a finding related to students' ability to think about their thinking:

A "metacognitive" approach to instruction can help students learn to take control of their own learning by defining learning goals and monitoring their progress in achieving them. (Donovan, Bransford, and Pellegrino 1999, p. 13)

Research on the performance of experts suggests that they reflect on and monitor their understanding of a scientific investigation. Experts note any requirement for additional information, the alignment of new information with what is known, and the use of analogies that may provide insights and advance their understanding. For experts, there often are internal conversations grounded in the processes of scientific inquiry. This finding has clear implications for the theme of teaching science as inquiry.

These three findings tell us that a science curriculum should acknowledge the fact that students already have ideas about objects, organisms, and phenomena, and that many of these ideas do not align with contemporary scientific understanding. The challenge for curriculum developers and teachers is not so much the fact that students have these misconceptions, but how to change the current conceptions. In contrast to many contemporary programs, research on learning indicates that curriculum and instruction should include a clear conceptual framework as well as facts and information. Finally, students can enhance their own learning through self-reflection and self-monitoring. These can be taught in the context of school science programs.

Whether carefully designed by professional curriculum developers or quickly compiled by teachers, any science curriculum provides an answer to the question "What should students know and be able to do?" Based on research, the general answer to the question seems clear: Students should learn both facts and concepts. But—and this is an important qualification—the curriculum should be structured using a conceptual foundation. The goal of learning with understanding includes factual knowledge placed on a conceptual framework. I underscore the complementary nature of these two ideas because many contemporary programs and assessments give much greater emphasis to facts without attention to the underlying concepts. Do we emphasize "big ideas" or "many facts"? The answer to this question is that we include both.

A science curriculum should be structured using a framework that includes ideas central to disciplines, such as the structure and properties of matter, biological evolution, and geochemical cycles for the physical, life, and Earth sciences, respectively. In addition, students should learn about science; that is, ideas fundamental to the process of science should be part of the curriculum. The content of school science should include scientific inquiry and the nature of science and fundamental ideas such as the empirical nature of science and the role of evidence in scientific explanations. Finally, and very important, the curriculums include opportunities for students to apply their understandings to personal and social issues. Table 1.1 presents a contemporary example of a framework for a high school science curriculum. The framework is from the BSCS program *BSCS Science: An Inquiry Approach* (2005).

What Should Students Be Able to Do?

The reader likely has noted that this discussion only answers part of the question "What should students know and be able to do?" Still to be addressed is the issue of the abilities that should be developed in a school science curriculum based on the National Science Education Standards (NRC 1996). Here are my nominations that address this issue:

- Identify questions and concepts that guide scientific investigations.
- Think critically and logically to make relationships between evidence and explanations.
- Formulate and revise scientific explanations and models using logic and evidence.
- Recognize and analyze alternative explanations and predictions.
- Communicate and defend a scientific argument.

Of course, these abilities will have to be developed in the context of investigations, experiments, and personal and social situations.

By this point, it should be clear that designing science curriculum is a complex process. This discussion only scratched the surface of the complex, interrelated issues that must be addressed. A coherent and rigorous science curriculum consists of a conceptual framework that includes major scientific ideas that have been sequenced with a clear progression for the concepts, the incorporation of scientific inquiry as content, and contexts such as personal and social perspectives.

Table 1.1

BSCS Science: An Inquiry Approach Framework for Grades 9–11

Units	Major Concepts Addressed at Each Grade Level		
	Grade 9	Grade 10	Grade 11
Science as Inquiry	Questions and concepts that guide scientific investigations	Design of scientific investigations Communicating scientific results	Evidence as the basis for explanations and models Alternative explanations and models
Physical Science	Structure and properties of matter Structure of atoms Integrating chapter	Motions and forces Chemical reactions Integrating chapter	Interactions of energy and matter Conservation of energy and increase in disorder Integrating chapter
Life Science	Cell structure and function Behavior of organisms Integrating chapter	Biological evolution Molecular basis of heredity Integrating chapter	Matter, energy, and organization of living systems Interdependence of organisms Integrating chapter
Earth-Space Science	Origin and evolution of the universe Origin and evolution of the Earth system Integrating chapter	Geochemical cycles Integrating chapter	Energy in the Earth system Integrating chapter
Science in a Personal and Social Perspective; Science and Technology	Personal and community health Natural and human-induced hazards Abilities of technological design	Population growth Natural resources Environmental quality	Science and technology in local, national and global challenges Understandings about science and technology

Research on Teaching

Although well-designed and sequential instructional materials can enhance student learning, they cannot do it all. The interaction between teachers and students is the key. Incorporating an instructional model into the instructional materials brings the materials as close as possible to facilitating the best interaction between teachers and students.

The BSCS 5E Instructional Model

The five components of what is referred to as the BSCS 5E Instructional Model are Engage, Explore, Explain, Elaborate, and Evaluate. The authors of *How People Learn* recommended an instructional sequence very close to the 5E Model. I quote from *How People Learn* (Bransford, Brown, and Cocking 1999, p. 127; emphasis mine):

> *An alternative to simply progressing through a series of exercises that derive from a scope and sequence chart is to expose students to the major features of a subject domain as they arise naturally in problem situations. Activities can be structured so that students are able to* explore, explain, extend, *and* evaluate *their progress.* Ideas are best introduced when students see a need or a reason for their use—*this helps them see relevant uses of knowledge to make sense of what they are learning.*

This summary, based on research, supports an instructional sequence very similar to the 5E Model. Table 1.2 displays the 5E Model. I have discussed the origin and use of the 5E Model elsewhere (Bybee 1997). In addition, my colleagues and I completed a review of research on the BSCS 5E Instructional Model (Bybee et al. 2006).

Table 1.2

BSCS 5E Instructional Model

Engage	The instructor assesses the learners' prior knowledge and helps them become engaged in a new concept by reading a vignette, posing questions, presenting a discrepant event, showing a video clip, or conducting some other short activity that promotes curiosity and elicits prior knowledge (Champagne 1987).
Explore	Learners work in collaborative teams to complete lab activities that help them use prior knowledge to generate ideas, explore questions and possibilities, and design and conduct a preliminary inquiry (Renner, Abraham, and Bernie 1988).
Explain	To explain their understanding of the concept, learners may make presentations, share ideas with one another, review current scientific explanations and compare these to their own understanding, or listen to an explanation from the teacher that guides the learners toward a more in-depth understanding (Renner, Abraham, and Bernie 1988).
Elaborate	Learners elaborate their understanding of the concept by conducting additional lab activities. They may revisit an earlier lab and build on it or conduct an activity that requires an application of the concept (Renner, Abraham, and Bernie 1988).
Evaluate	The evaluation phase helps both learners and instructors assess how well the learners understand the concept and whether or not they have met the learning outcomes (Kulm and Malcom 1991).

Source: Biological Sciences Curriculum Study (BSCS), the SCI Center. 2002. *Profiles in science: A guide to NSF-funded high school instructional materials.* Colorado Springs, CO: BSCS. p. 45.

Cognitive research findings indicate that learning is an active process occurring within and influenced by the learner. Hence, learning outcomes are an interactive result of what information is encountered and how the student processes that information based on perceived notions and existing personal knowledge. The BSCS 5E Instructional Model is an application of this research to curriculum materials and professional development experiences.

In *How People Learn*, the authors synthesized key ideas about learning based on an exhaustive review of the related research and identified parallel implications for classroom instruction (Bransford, Brown, and Cocking 2000). It is clear from both the analysis of curriculum and instruction in the Trends in International Mathematics and Science Study (TIMSS) project (Schmidt et al. 1999, 2001) and the work of the American Association for the Advancement of Science (2001) that these research-based ideas for instruction are neither commonly practiced in U.S. classrooms nor well supported in curriculum materials. Table 1.3 provides an overview of how the key findings are addressed in research-based curriculum materials (Powell, Short, and Landes 2002).

Table 1.3

Key Findings About Learning and Teaching and a Curriculum Response

Key Findings: Students	Key Findings: Teachers	Key Findings: Materials
Come to class with preconceptions	Recognize preconceptions and adjust instruction	Include structured strategies to elicit and challenge student preconceptions
		Incorporate background for the teacher about common preconceptions
Need to develop a deep factual understanding based in a conceptual framework	Understand content and conceptual framework. Provide examples for content	Are organized around a conceptual framework. Connect factual information to the framework. Provide examples for key ideas
Set goals and analyze progress toward them	Provide time for goal setting and analysis. Teach metacognitive skills	Make learning goals explicit. Integrate metacognitive skill development into content

The Instructional Model and Professional Development

In addition to the studies documenting student learning, other studies indicate that the 5E Instructional Model is perceived as valuable for professional development and for helping teachers reconsider how they teach. When interviewed and surveyed, teachers indicated that use of the 5E Model as an explicit instructional model has changed the way they think about student learning and, therefore, influenced their teaching. This response holds for teachers using curriculum materials based on the 5E Model and for those who have learned the model in a professional development session without specific materials (Lamb 2001, 2002a, 2002b; Funk 2002).

Although the three principles of how students learn science have a research base and clear implications for practice, they are more general than required for classroom science teachers. It also should be clear that the principles apply directly to the instructional core. The challenge for science teachers is to identify instructional materials that include three elements: a framework of scientific concepts and complementary facts; an instructional sequence that engages prior understandings and provides opportunities for students to develop new understandings and meanings through multiple and varied experiences; and, finally, assessments that provide feedback to the student and teacher about the degree to which learning has occurred.

Integrated Instructional Units and the Laboratory

America's Lab Report: Investigations in High School Science (NRC 2006) reported the effectiveness of laboratory experiences. The committee proposed using the phrase *integrated instructional units* to describe sequences of instruction that connect laboratory investigations with other types of teaching strategies such as reading, discussion, lectures, and web searches. Using a framework based on scientific inquiry, students might be engaged by framing research questions, making observations, designing an investigation, gathering data, and using those data to construct an explanation. Descriptions of integrated instructional units are closely related to the BSCS 5E Instructional Model. Sequences of laboratory experiences combined with other forms of instruction show this approach is effective for achieving three goals: improving mastery of subject matter, developing scientific reasoning, and cultivating interest in science. Furthermore, integrated instructional units appear to be effective in helping diverse groups of students make progress toward these goals. See Table 1.4 for a summary of research-based conclusions. What follows is a major conclusion from this NRC report.

> *Four principles of instructional design can help laboratory experiences achieve their intended learning goals if: (1) they are designed with clear learning outcomes in mind, (2) they are thoughtfully sequenced into the flow of classroom science instruction, (3) they are designed to integrate learning of science content with learning about the processes of science, and (4) they incorporate ongoing student reflection and discussion. (NRC 2006, p. 6)*

In conclusion, this discussion implies that design of curricular materials will enhance learning if it (1) applies the Understanding by Design (Wiggins and McTighe 2005), (2) uses an instructional model to sequence the student experiences, (3) explicitly integrates science concepts and inquiry abilities and understandings, and (4) provides ample opportunities for students to work in structured groups where they reflect on and can discuss their work.

Table 1.4

Attainment of Educational Goals in Typical Laboratory Experiences and Integrated Instructional Units

Goal	Typical Laboratory Experiences	Integrated Instruction Units
Mastery of subject matter	No better or worse than other modes of instruction	Increased mastery compared with other modes of instruction
Scientific reasoning	Aids development of some aspects	Aids development of sophisticated aspects
Understanding of the nature of science	Some evidence of increased interest	Some improvement when explicitly targeted at the goal
Understanding the complexity and ambiguity of empirical work	Inadequate evidence	Inadequate evidence
Development of practical skills	Inadequate evidence	Inadequate evidence
Development of teamwork skills	Inadequate evidence	Inadequate evidence

Teaching Science as Inquiry: Teaching Both Content and Procedures

Leaders in science education have the obligation to clarify a basic confusion that persists regarding scientific inquiry as it applies to education programs and to confront the controversial view that an inquiry orientation lacks intellectual rigor. Critics often reduce teaching science as inquiry to its simplest and most inappropriate form and summarily dismiss both the content and process. Unfortunately, inquiry has become associated with an ambiguous instructional approach and often is not recognized as a viable and appropriate set of educational outcomes, namely, the cognitive abilities and conceptual and factual understandings aligned with this central feature of the scientific enterprise. One hears arguments that inquiry approaches (note that use of terms such as *approaches* and *strategies* assumes that inquiry refers to teaching methods) are not effective for learning all science content because the process takes too long. The term is misinterpreted, extended to its most unreasonable position, and dismissed as not viable. Science teachers need to participate in efforts to clarify what the education community means by scientific inquiry: It is a content goal—that is, students should understand scientific inquiry and develop cognitive abilities. Inquiry also can be instructional approaches to achieve these goals.

Undoubtedly, some confusion about teaching science as inquiry emerges from the fact that inquiry is both a set of instructional strategies (e.g., laboratory investigations and activities) and education outcomes (e.g., knowledge such as "science advances through legitimate skepticism" and abilities such as "thinking critically and logically to make relationships between evidence and explanations").

A Brief History of Inquiry

Inquiry has been an explicit goal of science education for almost 50 years (Bybee and DeBoer 1993). In *A History of Ideas in Science Education*, George DeBoer states, "If a single word had to be chosen to describe the goals of science educators during the 30-year period that began in the 1950s, it would have to be inquiry" (1991, p. 206). Like many goals, inquiry provides a rallying point of apparent common agreement that fosters a sense of community and support among the advocates. Also, as is common with most education goals, there emerges the need for concrete examples of the abstract ideas and attitudes conveyed by the goal. This discussion provides examples of inquiry in the science curriculum as one approach to making the abstract more concrete in science education.

An Example of Inquiry in Curriculum and Instruction

From its earliest days, BSCS had included inquiry in its programs. Indeed, in the late 1950s, the deliberate, explicit, and comprehensive inclusion of "biology as inquiry" was a radical departure from other biology textbooks. At the time, H. J. Muller, a Nobel laureate and BSCS steering committee member, stated, "The trouble is not that there is too much science but too much shortsighted application of it, too little dissemination of its deeper meanings, and too little appreciation of the need for proceeding by its method of free inquiry" (1957, p. 252). Although scientists such as Bentley Glass, H. J. Muller, Bruce Wallace, and John Moore supported the inclusion of inquiry, Joseph Schwab likely contributed the most to actually implementing the theme of science as inquiry. Schwab's classic statement on the theme, his 1961 Inglis Lecture titled "The Teaching of Science as Enquiry," became a foundational statement for curriculum development (Schwab 1966).

Inquiry in Textbooks and Laboratories

The original BSCS programs used four avenues for implementing inquiry. First, the texts used expressions that indicated the uncertainty and incompleteness of science and the possibilities that through inquiry the uncertainty might be reduced and the knowledge made more complete. The texts had phrases such as *scientists are uncertain, we have been unable to discover the mechanism,* and *the favored theory is....* The phrases were designed to provide an accurate view of science and help students understand science as inquiry. Second, BSCS programs tried to replace a "rhetoric of conclusions" with a "narrative of inquiry." The texts included discussion indicating that science advances stepwise through investigations, data, and interpretations of data. Third, laboratory work was organized so it conveyed the sense that science, as the students experienced it, was inquiry. Although many laboratories were traditional—that is, designed to help students understand a concept—some were truly investigatory. Students investigated questions for which the text did not provide an answer. Some laboratories in the texts, the supplemental laboratory blocks, and the research problem series were examples of this mode of inquiry in BSCS programs. Finally, there were "Invitations to Inquiry" that provided another means to implement science as inquiry in biology programs (Schwab 1963).

For the most part, BSCS implemented inquiry in biology courses through the laboratories that accompanied the textbooks. With time, the market's influence on the revision of BSCS textbooks and declining support for the innovative NSF programs resulted in wider acceptance of conventional textbooks and decreased implementation of BSCS programs. Teaching science as inquiry became associated with doing laboratories, the primary aim of which was learning conceptual principles of science. By the early 1980s, adoption of BSCS programs was at a low ebb. The science teaching profession had lost sight of Joseph Schwab's rich explanations of the inquiry theme. Science textbooks had moved inquiry to the background, and science teachers began equating inquiry with hands-on approaches. To state the situation directly, the goal of inquiry had been reduced to a few laboratories and a slogan.

Ironically, it also was in the 1980s that research supporting the efficacy of teaching science as inquiry, and especially the effectiveness of BSCS programs, began emerging (Shymansky, Kyle, and Alport 1983; Shymansky, Hedges, and Woodworth 1990).

The National Science Education Standards

Publication of the *National Science Education Standards* in 1996 gave new life to the inquiry goal. The Standards presented inquiry as a prominent theme for both teaching and content. Here is a quotation from that document:

> *Scientific inquiry refers to the diverse ways in which scientists study the natural world and propose explanations based on the evidence derived from their work. Inquiry also refers to the activities of students in which they develop knowledge and understanding of scientific ideas as well as an understanding of how scientists study the natural world.* (NRC 1996, p. 28)

The actual standard states, "As a result of activities in grades 9–12, all students should develop both abilities necessary to do scientific inquiry and understandings about scientific inquiry." Figures 1.2 and 1.3 present the fundamental abilities and understandings associated with these respective components of the Standards.

Inquiry as Both Content and Process

The Standards shifted the implementation of the "science as inquiry" theme from an emphasis on the processes to a focus on cognitive abilities such as reasoning with data, constructing an argument, and making a logically coherent explanation (see Figure 1.2). Furthermore, the Standards made it clear that the aims of science education include students' understanding inquiry (see Figure 1.3).

As the science education community considers the needs at the instructional core—curriculum, professional development, and research and evaluation—we must acknowledge that our most valuable work occurs at the intersection of these three spheres. These questions are central: Given our mission of leadership and our goal of science literacy for all, the needs of students and teachers, and the current wave of assessments that grip the nation, what should the curriculum materials look like? What should our professional development programs provide? And how can we structure our research and evaluation to provide the evidence that schools and districts need when making decisions?

Discussions and subsequent decisions should be informed by a number of recent works: *National Science Education Standards* (NRC 1996); *How People Learn* (Bransford, Brown, and Cocking 1999); results of TIMSS (Schmidt et al. 2001); *Understanding by Design* (Wiggins and McTighe 2005); and recent evaluations by Project 2061 at the American Association for the Advancement of Science (AAAS 1999, 2000; Bybee 2001).

Figure 1.2

Abilities of Scientific Inquiry

- Identify questions and concepts that guide scientific investigations
- Design and conduct scientific investigations
- Use technology and mathematics to improve investigations and communications
- Formulate and revise scientific explanations and models using logic and evidence
- Recognize and analyze alternative explanations and models
- Communicate and defend a scientific argument

Figure 1.3

Understandings About Scientific Inquiry

- Scientists usually inquire about how physical, living, or designed systems function.
- Conceptual principles and knowledge guide scientific inquires.
- Scientists conduct investigations for a wide variety of reasons.
- Scientists rely on technology to enhance the gathering and manipulation of data.
- Mathematics is essential in scientific inquiry.
- Scientific explanations must adhere to criteria such as the following: A proposed explanation must be logically consistent, abide by the rules of evidence, be open to questions and possible modification, and be based on historical and current scientific knowledge.
- Results of scientific inquiry—new knowledge and methods—emerge from different types of investigations and public communication among scientists. In communicating and defending the results of scientific inquiry, arguments must be logical and demonstrate connections between natural phenomena, investigations, and the historical body of scientific knowledge. In addition, the methods and procedures that scientists used to obtain evidence must be closely reported to enhance opportunities for further investigation.

Inquiry and Assessment

Right behind the common standards movement is a related increase in state-wide testing. Federal and state agencies began to ask for evidence that students were meeting the standards that they had set; they began to ask for evidence of learning. Consequently, curriculum materials must provide students with opportunities for learning that could translate into higher achievement, such as test scores. Although materials could not and will not cover all the specific concepts and content included on every state test in science, curriculum developed in line with the standards will provide a solid foundation of knowledge that should serve students well on any well-conceived and well-developed assessment. Instructional materials that focus on content as articulated in the standards, present that content in a coherent framework, promote critical-thinking skills, give students exposure to the major concepts in science, and provide them with strategies for organizing and monitoring their own learning should help students perform better on local, state, national, and international assessments. Also of importance is the awareness that such testing in science covers all the disciplines of science, not just the life sciences, which in the recent past has been the discipline of science to which most high schools students have exposure (Blank and Langesen 2001).

A valuable resource for leaders is *Inquiry and the National Science Education Standards* (NRC 2000). This guide for teaching and learning contains discussions of inquiry as it is described in the standards and applied in classrooms. It includes clarifying examples of inquiry in science and assessments of inquiry in classrooms and makes the case for inquiry by making connections to our knowledge about how students learn. Another important aspect is that it describes essential features of classroom inquiry (see Table 1.5).

Table 1.5

Essential Features of Classroom Inquiry and Their Variations Along a Continuum

More<-----------------------Amount of Learner Self-Direction----------------------->Less Less<-----------------Amount of Direction From Teacher or Written Material----------------->More				
Learner ENGAGES in scientifically oriented questions.	Learner poses a question.	Learner selects among questions, poses new questions.	Learner sharpens or clarifies question provided by teachers, materials, or other source.	Learner engages in question provided by teacher, materials, or other source.
Learner gives priority to EVIDENCE in responding to questions.	Learner determines what constitutes evidence and collects it.	Learner directed to collect certain data.	Learner given data and asked to analyze.	Learner given data and told how to analyze.
Learner formulates EXPLANATIONS from evidence.	Learner formulates explanation after summarizing evidence.	Learner guided in process of formulating explanations from evidence.	Learner given possible ways to use evidence to formulate explanation.	Learner provided with evidence.
Learner connects explanations to scientific KNOWLEDGE.	Learner independently examines other resources and forms the links to explanations.	Learner directed toward areas and sources of scientific knowledge.	Learner given possible connections.	Learner given connections to scientific knowledge.
Learner COMMUNICATES AND JUSTIFIES explanations.	Learner forms reasonable and logical argument to communicate explanations.	Learner coached in development of communication.	Learner provided broad guidelines to use to sharpen communication.	Learner given steps and procedures for communication.

Source: National Research Council (NRC). 2000. *Inquiry and the national science education standards: A guide for teaching and learning.* Washington, DC: National Academies Press. p. 29.

In conclusion, scientific inquiry has a long, rich, and appropriate place in school programs. Science teachers can begin by applying a general understanding from the instructional core—namely, how teachers understand the nature of scientific knowledge—and enhance that understanding by applying the principles of learning discussion from the previous section. The obvious extension is how to apply appropriately the ideas about scientific inquiry to teaching and classwork. I have discussed various aspects of scientific inquiry in prior works (Bybee 1997, 2002, 2005).

Professional Development: Improving Science Teachers' Knowledge and Skills

Everything I have discussed—scientific literacy, incorporating research, and an inquiry orientation—requires some level of professional development for those in the science education community, and that is the personal obligation of science teachers and the professional responsibility of education leaders. Professional development is among the most significant trends in education in the first decades of the 21st century. One could view the categories I have presented as topics or themes within which to initiate professional development programs.

Professional development experiences should not be instituted as single "events" with topics that may be interesting but nonetheless isolated from the central work of science teaching. So, for example, I recommend that professional development be seen as integral to science curriculum reform in school systems. Assessment certainly could also be a central entry point for professional development.

National Standards and Professional Development

It is probably worth recalling the professional development standards because they serve as an initial orientation for leaders. Here are the essential statements on professional development from the *National Science Education Standards* (NRC 1996):

- Professional development for teachers of science requires learning essential science content through the perspectives and methods of inquiry (p. 59).
- Professional development for teachers of science requires integrating knowledge of science, learning, pedagogy, and students; it also requires teachers to apply that knowledge to science teaching (p. 62).
- Professional development for teachers of science requires building understanding and ability for lifelong learning (p. 68).
- Professional development programs for teachers of science must be coherent and integrated (p. 70).

Standards for science education cannot change teachers' beliefs or behaviors, but they can provide clear and crucial directions for change because they define goals and identify directions for improvement. Standards do have the power to change elements at the instructional core and provide a vision of what should be maintained and what should be changed within science education.

When developing the national standards, we recognized professional development as a key component of the science education system, one often neglected when concentrating on core content. The professional development standards consider *what* teachers should learn and *how* they should learn it. Although the contexts for professional development will vary, leaders can use the standards as a model for designing professional development. Note that the terms integrating and integrated are used in two of the four standards, thus reinforcing the idea that professional development should be seen as central to leadership activities, especially as it applies to science teachers.

Professional Development and the Curriculum

One area of support—teacher professional development and its critical link to curriculum—seems essential. Researchers and practitioners have identified two important ways in which new curriculum and professional development are symbiotic. First, for teachers to use new curricula well, especially those materials that incorporate new content and teaching requirements, they need opportunities to learn new knowledge, skills, and approaches to instruction. Professional development is required for new curricula to be used well and with fidelity.

Although that idea is far from new, another research finding is a more recent discovery. New curriculum materials appear to be effective vehicles for teacher learning. By studying new materials, using them in classrooms, examining the thinking and products of students who interact with the materials, and sharing their observations and dilemmas with others, teachers can strengthen their understanding of content, student learning, and effective teaching strategies (Ball 1996; Cohen and Hill 1998; Russell 1998). Professional development based on and using an integrated science curriculum can provide students with a variety of ways to learn that focus on fundamental understandings and skills while also helping teachers learn more content and more effective ways to teach.

The book *Designing Professional Development for Teachers of Science and Mathematics* (Loucks-Horsley et al. 2003) describes 15 strategies that are being used successfully to help teachers learn. Three of these strategies link to curriculum. One of these strategies, curriculum development, can help teachers learn in situations when appropriate curriculum materials do not exist or when teachers have the expertise, time, and resources necessary to develop their own. Another strategy, the use of replacement units, helps teachers learn how to teach in a new way, or teach content that is new to their curriculum, by trying a series of lessons and then reflecting on their experiences. This strategy is especially helpful when no full set of curriculum materials exists, when a small foray into a new approach to teaching and learning is called for, or when units are being purchased and introduced one at a time for budgetary reasons. The third strategy, curriculum implementation, is most relevant when discussing the use of a full set of curriculum materials such as an integrated science program.

Implementing New Curriculum

Curriculum implementation uses different kinds of professional development as teachers' concerns, knowledge, and skills change. Early in implementation, professional development through a workshop format can help teachers learn "how to do it" from those experienced in using the curriculum. By experiencing the student activities as a learner, teachers learn the content they will teach their students. For some who learned science as a set of facts to be memorized, this method helps them learn what a "big idea" (fundamental science concept) is and how facts fit into conceptual understanding. Science teachers also may experience an inquiry-based approach. They learn to develop explanations for what they observe and defend their explanations using evidence. They learn to challenge each other's ideas in an educational context and report on their own using effective methods of communication.

In effective professional development, teachers learn to wear two hats: a hat of a learner and that of a teacher. They practice doing science through the student activities and reflecting on what and how they are learning. Then they learn to step back and view their experiences as a teacher. What ideas did they struggle with learning? What helped them in their struggles? What ideas will their students struggle to learn? Discussions of questions such as these go beyond management to concerns and the consequences—student learning—of new science programs.

These activities in initial workshops will prime teachers to teach the science curriculum. In the most supportive settings, they have coaches who visit and look for ways to help them make their classroom and material management more efficient. They have opportunities to share their successes and problem solve with other teachers, including those experienced in using the curriculum.

Professional development changes when management concerns decrease and teachers can focus on instructional practice. Strategies for teacher learning include examininations of student work, case discussions, and action research (Loucks-Horsley et al. 2003). Teachers can share student work that raises questions for them: How can I help my students realize the role of inquiry in the study of science? What are the current conceptions of these students? Other issues that student work may illustrate include the development of inquiry abilities and the use of technology in science and in school programs. Similarly, teachers can view a video from one of their classrooms or a video collection, write and share teaching cases, or read cases that have been developed by others, and talk about student learning and teaching issues that arise in these activities. The use of both video and narrative cases is increasing as evidence of their value for teaching and learning accumulates (Barnett 1998; Schifter 1996).

Student Achievement: Attaining Higher Levels for All Students

Since the 1960s, the United States has been part of the tradition of international comparative studies of mathematics and science education. The results of two assessments, PISA (Programme for International Student Assessment 2003; Lemke et al. 2004) and TIMSS (Mullis et al. 2001; Gonzales, Guzman et al. 2004; Gonzales, Pahlke et al. 2004), have engaged our interest. The primary domain for PISA 2006 was science, which should affect the work of science education leaders (Bybee and McCrae 2009). At some times and in some categories, the United States does better than other countries. But more often than not, we are about average. Occasionally, we are among the lowest-achieving of all countries in this assessment. The results can be reviewed in *Pursuing Excellence: Eighth-Grade Mathematics and Science Achievement in the United States and Other Countries From the Trends in International Mathematics Science Study* (Gonzales et al. 2004), *International Outcomes of Learning in Mathematics Literacy and Problem Solving: PISA 2003 Results From the U.S. Perspective* (Lemke et al. 2004), and *PISA Science 2006: Implications for Science Teachers and Teaching* (Bybee and McCrae 2009).

An Apparent Contradiction

Beginning in the 1990s, I became intrigued with an apparent contradiction that accompanies these international assessments. Inevitably, we compare our results—as one nation—to other countries', yet states and school districts decide what students should know and what curriculum to implement.

With the release of national comparative results, the inevitable commentaries express varying degrees of concern. If the United States does as well compared to other countries as we did at the fourth-grade level on the 1995 TIMSS, we hear comments that we really should have done better. After all, we should be first in the world. If we do as poorly as our physics and advanced mathematics students did on the Third International Math and Science Study—no countries performed more poorly, that is, we are last in the world—then commentaries express deep concern because these students represent our best and brightest.

Now, let us examine the contradictions in detail. With international assessments we compare the United States to other nations, and we do this assuming the U.S. results represent one nation. At the same time, we maintain an education system that defends the right of 50 states and the District of Columbia, and approximately 15,000 local districts, to make decisions about what students should know and be able to do, how they should be taught, and what content and processes should be assessed. Here we see the two contradictory aspects of the situation. We value a system with wide and significant variation but appeal to assessment results as though we were one unified system.

How might this contradiction be resolved? This basic challenge centers on maintaining the rights of states and local jurisdictions to determine the curriculum, instruction, and assessments and, at the same time, attaining higher student achievement as a nation. This suggests a place for national common core standards for science education.

No Child Left Behind

The No Child Left Behind Act of 2001 (NCLB) must be considered because it has been a dominating influence on science education, and its implementation will continue after the introduction of assessment for science in the 2007–08 school year. The NCLB uses assessment results as a punitive means to ensure that schools make adequate yearly progress in student achievement. To meet this goal, the federal law requires states to set high standards, ensure highly qualified teachers, and implement yearly assessments—all at the state level. But note that states are still setting the standards and implementing the assessments. Most financial support for changes designed to accommodate the NCLB mandates goes directly to the states, so NCLB avoids establishing a national curricula, and I argue that it holds little promise of attaining higher levels of student achievement—as one nation. Because, by design, NCLB yields decisions about standards, curriculum, instruction, and assessment to the states, it does not avoid the fundamental causes of incoherence at the core of the education system.

A Proposed Solution

I can propose a resolution to the problem. The science framework for the 2009 National Assessment of Education Progress (NAEP) and common core standards hold promise both for maintaining the rights of states and school districts to select instructional program and for attaining higher levels of achievement—as a nation.

First, it is important to understand that use of, for example, the NAEP framework and standards such as the *National Science Education Standards* (NRC 1996) and common core standards is voluntary.

Second, the framework standards define and describe what students should know and be able to do. The NAEP framework and national standards include the science understanding and abilities students should develop as a result of their K–12 education. They do not prescribe the structure, organization, balance, or presentation of content and processes in classrooms. To be clear, national standards and assessment frameworks are not lessons, classes, courses of study, or school science programs. This said, they do have the capacity to influence core components of the education system—namely, curriculum, instruction, classroom assessments, and complementary aspects of the system such as teacher education and initial certification and continued professional development of teachers and administrators. National standards and the 2009 NAEP framework can serve as a weak force field that steadily influences decisions over time about state standards, adoption requirements, textbooks, teacher education programs, and state and local assessments. Greater alignment among core components of the education system will enhance student learning and result in higher levels of achievement as a nation. So the 2009 NAEP framework and national standards may resolve the apparent contradiction by the nature of their influence on state and local systems, even as they contribute to greater consistency and coherence within the educational system.

Why Another Assessment—PISA?

With assessments such as NAEP and TIMSS, one can reasonably ask whether there is anything different about PISA. PISA does provide a unique perspective on the assessment landscape, one that complements other tests in that it specifically measures how well young adults are prepared to meet the challenges of today's scientific and technological world. The assessment focuses on young people's ability to use their knowledge and apply their skills to real-life situations. In contrast to assessments such as TIMSS and NAEP, PISA does not directly focus on curricular outcomes. PISA is not an assessment of the "attained curricula." PISA measures the application of knowledge in reading, mathematics, and science to problems in life situations. PISA scores represent the product or yield of learning experiences at age 15. This is the general picture of PISA. As an example of PISA's benefit, the 2006 assessment provides specific information that likely will be of interest to policy makers and educators as it pertains to the basic skills needed by U.S. students and our country's continuing interest in international competitiveness.

Students' Basic Skills

In the 2003 survey, PISA analyzed education and presented the basic skills students will need to thrive in a changing economy. We can take this list of skills and ask how U.S. students are doing compared with 15-year-olds in other countries, especially Organisation for Economic Co-operation and Development (OECD) countries, as they represent our main economic competitors. What are these basic skills? And what can we learn about U.S. students' performance on these skills from PISA 2003? The following paragraphs answer these questions.

The first basic skills should not surprise any science teacher: the ability to read and do math at the 9th-grade level or higher. The majority, 61%, of U.S. 15-year-olds participating in PISA were in 10th grade, and 30% were in 9th grade. It seems PISA would provide a good measure of U.S. achievement on these skills. In 2003, the average U.S. score in reading literacy was not measurably different from the OECD average. Nine countries outperformed the United States in reading literacy in that year's study. Between 2000 and 2003, there was no change in either the U.S. reading literacy score or the U.S. position compared with the OECD average.

U.S. 15-year-olds' mathematical literacy was below the OECD average and lower than 23 of the 38 countries whose data have been analyzed. PISA also reports percentages of students at six proficiency levels—low to high. Disappointingly, the United States had greater percentages at the lowest levels and fewer students at the higher levels of proficiency than the OECD average percentages. Half (19) of the 38 participating countries had more students at the highest level of proficiency (level 6) than the United States. These data indicate that the United States is holding its position in reading literacy and is behind most OECD countries in mathematical literacy.

Another basic skill involves the ability to solve problems. PISA included problem solving as a cross-disciplinary component of the 2003 survey. The PISA exercises assessed 15-year-olds' abilities to use reasoning processes to draw conclusions, make decisions, troubleshoot, and analyze procedures and structures of complex systems. The assessment required students to apply processes such as inductive and deductive reasoning, establishing cause-and-effect relationships, and combinatorial reasoning. Finally, the problem-solving assessment also related to other basic skills, namely, working toward solutions and communicating the solution to others through appropriate representation. U.S. students scored lower than their peers in 25 of the 38 countries. In short, our students' performance on problem solving does not compare well with other industrialized democracies.

U.S. students are not doing as well as expected in several basic skills that are essential to individuals' potential in the workforce and our nation's economic development. Given this disappointing performance, one might reasonably ask about the expectations of U.S. 15-year-olds as far as their future education and occupations. Well, PISA did ask about these goals. Sixty-four percent of U.S. students reported that they expected to complete a bachelor's degree or higher. This was much higher than the OECD average (44%). Only South Korea reported a higher percentage than the United States. U.S. students who expected to complete a bachelor's degree did score higher than their peers with lower educational expectations. However, compared with the 37 other countries reporting data, U.S. students were outperformed by their peers in 26 countries in mathematics literacy.

Students were also asked about their job expectations by age 30. Responses were coded according to one of the groups in the International Standard Classification of Occupations. The responses were then collapsed into three categories—high, medium, and low—based on skill level. Sixty-seven percent of U.S. students reported high job expectations; 32% reported medium expectations. The OECD average for high occupational expectations was 47%. As with degree expectations, U.S. students with high job expectations scored higher on the mathematics literacy scale than their U.S. peers with lower expectations. However, they scored lower than the OECD average in mathematics literacy for students with high job expectations.

These data showing comparably low scores on reading, mathematics, and problem solving contrasted with high educational and occupational expectations should be cause for concern, if not alarm. We have to not only increase performance in reading, mathematics, problem solving, and (I would add) science but also establish accurate and reasonable expectations for education and careers.

Perhaps the most educationally significant insight to be gained from PISA emerges from the difference between TIMSS and PISA. The difference I refer to is the orientation or emphasis of respective assessment. TIMSS is grounded in the curriculum and provides feedback for how students are attaining what is intended and enacted vis-à-vis a country's curriculum. Although it does not ignore school curriculum, PISA asks how well students can apply their knowledge in real-world situations. Lower scores on PISA suggest that our students do not do as well as the majority of our economic competitors when they have to demonstrate basic skills in solving contextual problems. This should be a concern for policy makers and educators alike.

The evidence indicates that our students perform reasonably well on curriculum-based assessments. (Some would question even this.) But our students do not perform very well on context-based assessments, especially those involving content knowledge and basic skills associated with economic productivity.

PISA provides a beneficial perspective, one that complements that of NAEP and TIMSS and that U.S. educators should take seriously when developing reviews and reforms of school curricula and instruction.

Concluding Discussion

In this chapter, I introduced the instructional core as the central focus of discussion in the book. Improving student learning at the instructional core involves raising the level of content that students are taught, increasing the skills and knowledge of teachers, and increasing the level of students' actual learning.

I also described five themes—achieving scientific literacy, reforming science programs, teaching science as inquiry, improving teachers' knowledge and skills through professional development, and attaining higher levels of student achievement—that all relate directly to both the instructional core and the unity of the topics presented in the following chapters.

The Teaching of Science Content

In this chapter, I present the ideals and spirit of Paul F-Brandwein. In particular, I bring contemporary ideas to themes that he presented almost 50 years ago. Those themes include the substance of science education, curricular structure, and the style of science teaching that emphasized inquiry as a fundamental aspect of science. Two of Brandwein's essays—"Elements in a Strategy for Teaching Science in the Elementary School" and "Substance, Structure, and Style in the Teaching of Science"—are the basis for the chapter. The former was the 1961 Burton Lecture presented at Harvard University, and the latter is based on several lectures given at different locations, including NSTA meetings (Brandwein 1962; Brandwein 1965). Consistent with these essays, I emphasize elementary school science as a focus, and I honor Paul's lifelong interest in the environment and conservation by using examples from the environmental sciences. By doing so, I build on this great leader's insights, ideals, and wisdom.

Brandwein's Original Themes: Substance, Structure, and Style

Brandwein began his monograph *Substance, Structure, and Style in the Teaching of Science* (Brandwein 1965) with a clarification of the knowledge products and investigative processes of science. Interestingly, he also made many explicit references to technologies. He pointed out that advances in science and technology affect our social, economic, and political lives. His main theme centered on the process of scientific literacy and the fact that "scientists start with a base: a conceptualization of the world as it appears to them" (Brandwein 1965, p. 2). He continued, "Children come to school with a base: a conceptualization of the world as it appears to them" (p. 2). Brandwein then makes an essential connection for this discussion.

> *And teachers meet children with a base: their conceptualizations, not only of the world as it appears to them, but of a boy and girl fitted to live in a world*

*changing daily because scientists exist and work. Every teacher brings …
substance, structure, and style.* (Brandwein 1965, p. 2)

The Substance of Science

Because the themes seem to be fundamental to science teachers, one has to ask,
What does Brandwein mean by substance, structure, and style? He elaborates on
the idea of substance by appealing to James Bryant Conant's definition: Substance
consists of a "series of concepts or conceptual schemes arising out of experiment and observation and leading to new experiments and ideas" (Conant 1957,
p. x). The conceptual schemes are not final statements. Rather, they are temporary scientific explanations.

> *They are the hard substances of science quarried by meaningful and relevant
> activity by the scientist in (a) investigating the material universe; (b) developing
> orderly explanations of the objects, events, and phenomena investigated; and (c)
> subjecting his orderly explanation, or concepts, to testing by insisting on and
> inventing means for empirical validation.* (Brandwein 1965, p. 3)

For Brandwein, substance was the fabric of ideas and tools of science and
technology. Adding dimension to the theme of substance, Brandwein weaves
a wonderful tapestry of scientific inquiry as a process leading to knowledge of
science and technology. As we will see, however, the substance of inquiry and
technology did not become a significant component of curricular structure.

The Structure of Curriculum

What about the theme of structure? For Brandwein, structure in science teaching
consists of how concepts are developed in the educational environment. The
structure includes content ordered in the forms of concepts and the educational
experiences that the students bring and that the science teacher provides. In
short, one can think of how concepts are organized and the relationship they
have to the students' experiences in and out of school. Of primary importance
for Brandwein was the relationship among concepts and the conceptual organization of the curriculum: "To learn structure, in short, is to learn how things
are related" (Brandwein 1965, p. 4). One should note that for Brandwein the
central organizing features of structure are the science concepts and how they
are related.

The Style of Science Teaching

Finally, style in science teaching is what the teacher does and the ways in which
science teachers accomplish their goals. Style includes what goes on in the
science classroom, including policies, procedures, and actual teaching practices.
Style includes the culture of the classroom and the norms of the learning envi-

ronment that the teacher establishes. Once the substance of science and a curricular structure have been established, the style of science teaching stems from the substance and structure, not the reverse. The idea just mentioned and the one that follows present important foundations for critical evaluations of contemporary science education. Brandwein directs our attention to inquiry, investigation, and problem solving but suggests that style represents more than the processes of science. For Brandwein, the scientists' art of investigation should have a role in the teaching of science.

In the following discussion, I use the terms *content*, *coherence*, and *congruence* to connect Brandwein's ideas of substance, structure, and style from the 1960s with 21st-century perspectives on those themes.

The Content of Science

Because of science, we have explanations for many of the objects, organisms, events, and phenomena around us. For example, we have explanatory theories about the particulate nature of matter; the genetic basis of heredity; the relationships among earthquakes, volcanoes, and plate tectonics; and the movements of objects in the solar system and beyond. This said, we also know that facts and information change, often quite rapidly. But concepts or conceptual schemes— statements of relationships, patterns of observed phenomena—remain relatively stable during long intervals of time.

As contemporary research summarized in *How People Learn* (Bransford, Brown, and Cocking 2000) indicates, students require conceptual frameworks within which to organize their continued learning. So, a recommendation for a conceptual foundation for school science programs has support from both science and learning theory. Brandwein proposed six conceptual schemes that could serve as the fundamental content of the science curriculum. The conceptual schemes Brandwein proposed are presented in Figure 2.1.

Figure 2.1

Conceptual Schemes for the Science Curriculum

- When energy changes from one form to another, the total amount of energy remains unchanged.
- When matter changes from one form to another, the total amount of matter remains unchanged.
- The universe is in continuous change.
- Living things are interdependent with one another and with their environment.
- A living thing is the product of its heredity and environment.
- Living things are in continuous change.

In an elaboration on the conceptual schemes, Brandwein pointed out that they are congruent with scientific theory, laws, and principles but not identical. Conceptual schemes help curriculum developers present scientific concepts in an effective manner to the students who will become our citizens. In the end, the test of a conceptual scheme is whether it helps students make a connection, an intellectual or cognitive link between scientific and technological concepts and citizens' explanations of meaningful phenomena. In school science, for example, one can ask about the degree to which the conceptual schemes provide a meaningful cognitive structure for student learning about the natural and designed world.

Content Standards and the Science Curriculum

The idea of conceptual schemes is present in the *National Science Education Standards* (NRC 1996) and will be in the proposed common core standards. The national standards represent one of the most significant contributions to science education in the latter part of the 20th century and perhaps in the history of science education in the United States.

The release of the national standards in late 1995 shifted the focus of educational discussion from the development of standards to their implementation through the curriculum. One can anticipate the same shift when new national standards for science education are developed and released. Educators begin asking questions about instructional materials that will help learners understand the science content defined in the standards. In several places, the standards point out that the categories used to present science content do not represent a curriculum or even a curriculum framework. Yet equating the content standards with a science curriculum persists as a prevalent misconception. In very broad strokes, the standards clearly have implications for the design of science curricula, but they do not propose a particular curricular organization or structure. As we shall see in the next sections, they have an orientation that ranges from concrete in the lower grades to abstract in the upper grades. This orientation acknowledges the need for developmentally appropriate statements of science content. Still, the essential curricular decisions remain with states and local educators. I quote from the standards:

> *Content is what students should learn. Curriculum is the way content is organized and emphasized; it includes structure, organization, balance, and presentation of content in the classroom. Although the structure for the content standards organizes the understandings and abilities to be acquired by all students K–12, that structure does not imply any particular organization for science curricula.* (NRC 1996, p. 111)

The science content described in the standards represents a conceptual level that is related to the conceptual schemes Paul Brandwein recommended, especially when you examine the fundamental understandings for a standard such as Organisms and Environments: At Grade K–4. The standards present science content that has the conceptual orientation Brandwein recommended, but the statements of content present the concepts at levels appropriate for grades K–4, 5–8, and 9–12. Indeed, any differences between Brandwein's conceptual sciences and the standards should contribute to greater implementation of the content by curriculum developers and, ultimately, by science teachers.

Conflicts Over the Content of Science Teaching

Release of the standards inevitably broadened and deepened discussions about science education in general and state and local standards in particular. Although the science education community had been aware of the standards' development and had many opportunities for review and input, the actual standards stimulated new discussions as different factions confronted the possibility of change. Such discussions are not new in the history of education or science education (Kliebard 1994).

Unfortunately, many debates about the content of science programs have neither recognized the different education goals and subtleties of curricular structure, nor acknowledged the historical contributions of individuals such as Paul Brandwein. Support continues to develop for the conceptual orientation originally presented by Brandwein and further expanded on by the national standards. Beyond his argument for a conceptual orientation for the curriculum, several other themes of Paul Brandwein's writing should be considered essential to the science curriculum. I am referring to content about scientific inquiry, technology as it relates to science, science as it connects to personal and social perspectives, and the history and nature of science. Content and the teaching of science include much more than memorizing facts and recalling information. For the teaching of science, we have obligations to provide all students with opportunities to develop an understanding of science and technology. Although my recommendations may seem reasonable and well supported by our history (DeBoer 1991; Bybee 1997) and the standards, conflicts still emerge over content. These conflicts can be characterized as an emphasis on facts and science content devoid of contexts versus an orientation such as the one recommended here. Furthermore, the conflicts often are played out in the political arenas of state standards, adoption of science textbooks, and budget priorities.

Here I provide a contemporary example of different perspectives on science content as it relates to school programs. I will quote extensive portions from two responses to an approach to curriculum reform as proposed by Marjorie G. Bardeen and Leon M. Lederman in the July 10, 1998 issue of *Science*. The statements point to different and conflicting positions of content for the teaching of science. The first quote is from the late John A. Moore, who was a member of the National Academy of Science and began his involvement in science education in the late 1950s. Dr. Moore was instrumental in BSCS programs, in particular the BSCS "Yellow Version."

No one knows what a different educational system appropriate for the nation's needs might be, and only experimentation with various patterns will indicate that programs in science education will better serve the nation as a whole and the students in the classrooms. One of the appalling defects to be overcome is simply this: almost without exception, there is no place in the kindergarten through grade 12 (K–12) system where students are provided a solid background of information that will enable them to make those sound decisions required by informed citizens in our complex society. In fact, no important human problem for which science and technology may be both a cause and solution is treated adequately. Some of these problems are the use of natural resources, health care, strengths and limitations of scientific procedures.

As far as these complex societal problems are concerned, K–16 education is largely irrelevant. The reason for this is that the students are presented with little more than the contents of one or several of the separate disciplines of science, but the critical step of using the information to consider human problems is rarely taken. Would it be more worthwhile to design the science curriculum with the goal of understanding both the natural and technological worlds that students experience? There must be an acceptance that science courses have to make that major step to relevance. There is quite a gap between understanding the chemistry of combustion and understanding how human societies will solve their needs for energy now and in the future. Students need to know both. (Bardeen and Lederman 1998, p. 178)

The second quote is from the late Glen Seaborg, who also was a member of the National Academy of Science and a Nobel laureate and who was involved in science education since the 1950s. Dr. Seaborg was instrumental in development of CHEM Study.

I agree that it is a necessary goal to strive for a scientifically literate society that can understand, even enjoy, the assimilation of scientific knowledge from the traditional disciplines, the interdependence of scientific discovery and technological advances, and the role of scientific knowledge and scientific ways of thinking that will be needed to address the major societal challenges facing the human race. However, I propose a pathway to this goal that is somewhat different from the one proposed by the authors, one that is more strongly discipline-based. Further, I would suggest that if all middle and high schools and their science departments are encouraged to follow the same sequence, the collective result will be improved instruction and a curriculum leading to higher levels of student achievement.

I take the position that the nature of modern science today can only be achieved if we prepare students in the fundamentals of the science disciplines that are most efficiently taught in a 6-year specific sequence in traditional, largely discipline-based courses. (Bardeen and Lederman mention this as an alternative approach.) Early introduction of the most central concepts is needed, and instruction should focus on the essential core content of science. After all, the many scientists who so enthusiastically endorse the approach called for by Bardeen and Lederman arrived at this "higher" conceptual perspective through traditional, discipline-based instruction. (Bardeen and Lederman 1998, p. 179)

These different perspectives can be and have been resolved in various ways over the years. Unfortunately, the conflicts over content have, in some cases, taken a course that is less than civil (Olson 1998; Bass 1998) and often relates to documents such as the standards (Metzenberg 1998). I mention the conflicts over the content of science teaching because they are realities of our age. In the early 21st century, science teachers must be prepared for conflicts by understanding the role and place of content, such as that proposed in the standards, and recognize the political aspects of curriculum reform.

Coherence and the Science Curriculum

Current discussions of coherence provide a contemporary perspective of Brandwein's theme of structure. To be specific, structure refers to a curriculum framework in which the major conceptual schemes were, for example, restated in seven levels generally aligned with grade levels in elementary school and progressed from kindergarten through sixth grade. This suggests that one level precedes the next so the students develop greater understanding. Brandwein developed structures for the elementary and junior high school levels and proposed that these curricular structures would provide a basis for comprehension of the more sophisticated conceptual schemes in Table 2.1 (pp. 36–37).

In the decades since Brandwein's discussions of curricular structure, many science programs have lost sight of the idea of a clear and consistent curricular structure based on conceptual schemes. Instead, we have curricular conglomerates based on a mix-and-match array of activities that lack conceptual coherence of clear progressions of learning outcomes. We now hear criticism of the curriculum based on a lack of coherence, and we hear about the results of incoherent science programs with each international assessment of student achievement (Schmidt et al. 2001).

Table 2.1

Conceptual Frameworks

	CONCEPTUAL SCHEME A	CONCEPTUAL SCHEME B	CONCEPTUAL SCHEME C
	When energy changes from one form to another, the total amount of energy remains unchanged.	When matter changes from one form to another, the total amount of matter remains unchanged.	The universe is in continuous change.
CONCEPT LEVEL VI	The amount of energy gotten out of a machine does not exceed the energy put into it.	In nuclear reactions, matter is converted to energy, but the total amount of matter and energy remains unchanged.	Nuclear reactions produce the radiant energy of stars and consequent change.
CONCEPT LEVEL V	Energy must be applied to produce an unbalanced force, which results in a change in motion.	In chemical or physical changes, the total amount of matter remains unchanged.	Bodies in space are in continuous change.
CONCEPT LEVEL IV	A loss or gain of energy affects molecular motion.	In chemical change, atoms react to produce change in the molecules.	The Earth's matter is in continuous change.
CONCEPT LEVEL III	The Sun is the Earth's chief source of radiant energy.	Matter consists of atoms and molecules.	There are seasonal and annual changes on Earth.
CONCEPT LEVEL II (Analogical)	Energy can change from one form to another.	A change in the state of matter is determined by molecular motion.	There are regular changes in positions of the Earth and Moon.
CONCEPT LEVEL I (Analogical)	Force is required to set an object in motion.	Matter commonly exists as solids, liquids, and gases.	There are daily changes on Earth.
BEGINNING CONCEPT LEVEL (Analogical)	A force is needed to start, stop, or change the direction of motion.	Matter is characterized by certain properties by which it can be identified and classified.	Things change (implicit within the development of Conceptual Schemes A and B).

Table 2.1 *(continued)*

Conceptual Frameworks

CONCEPTUAL SCHEME D	CONCEPTUAL SCHEME E	CONCEPTUAL SCHEME F	
Living things are interdependent with one another and with their environments.	A living thing is the product of its heredity and environment.	Living things are in continuous change.	
Living things depend basically on the capture of radiant energy by green plants.	Man is the product of his heredity and environment.	Changes in the genetic code result in changes in living things.	CONCEPT LEVEL VI
Living things are adapted by structure and function to their environment.	The cell is the unit of structure and function in living things.	Over the ages, living things have changed in their adaptation to the changing environment.	CONCEPT LEVEL V
Living things capture matter and energy from the environment and return them to the environment.	A living thing reproduces itself and develops in a given environment.	Living things are adapted to particular environments.	CONCEPT LEVEL IV
The Earth's different environments have their own characteristic life.	Living things are related through possession of common structure.	Living things grow and develop in different environments.	CONCEPT LEVEL III
Living things depend on their environment for the conditions of life.	Related living things reproduce in similar ways.	Forms of living things have become extinct.	CONCEPT LEVEL II (Analogical)
Living things are affected by their environment.	Living things reproduce their own kind.	There are different forms of living things.	CONCEPT LEVEL I (Analogical)
Environments differ (implicit within the development of Conceptual Scheme F).	Living things may differ in structure, but they have common needs and similar life activities.	Living things grow (implicit within the development of Conceptual Scheme E).	BEGINNING CONCEPT LEVEL (Analogical)

The Role of the 1996 National Standards and 21st-Century Common Core Standards

Increasing curricular coherence is one way to think about the power of national standards and changes they can effect in the science curriculum. Implementing standards has the potential to facilitate greater coherence among educational components. The assumption behind this position is that greater coherence among goals, curriculum, instruction, assessments, teacher education, and professional development will enhance students' achievement. By some reports—for example, the Third International Mathematics and Science Study (TIMSS)—we have an incoherent education system (Schmidt and McKnight 1998). Goals are only tangential to instructional materials, inconsistent with assessments, incongruent with professional development, and so on. I begin this discussion with a basic definition: Coherence occurs when a small number of basic components are defined in a system, organized in conceptual relationship to each other, and other components are based on or derived from those basic components. How will standards bring about greater coherence within science education? Over time, the standards for science education have the potential to develop coherence by

- defining the understandings and abilities of science that all students, without regard to background, future aspirations, or prior interest in science, should develop;
- articulating content, pedagogy, and assessments at different grade levels;
- coordinating programs for professional development; and
- providing criteria for evaluating current and proposed programs.

The Importance of Research on Learning

The National Research Council report *How People Learn: Brain, Mind, Experience, and School* (Bransford, Brown, and Cocking 1999) is a major synthesis of research on human learning. Findings from *How People Learn* have both a solid research base and clear implications for this discussion on curricular coherence. The following statement is from a subsequent report, *How People Learn: Bridging Research and Practice* (Donovan, Bransford, and Pellegrino 1999). The finding refers to the conceptual foundation of a curriculum.

> *To develop competence in an area of inquiry, students must: (a) have a deep foundation of factual knowledge, (b) understand facts and ideas in the context of a conceptual framework, and (c) organize knowledge in ways that facilitate retrieval and application.* (p. 12)

By transferring these recommendations to the curriculum and echoing Brandwein's recommendations from the 1960s, the developers of the science curriculum should incorporate fundamental knowledge that is based on, and contributes to, the students' development of a strong conceptual framework. Research comparing the performance of novices and experts, as well as research on learning and transfer, shows that experts draw on a richly structured information base. Although factual information is necessary, it is not sufficient. Essential to expertise is mastery of concepts that allow for deep understanding. Such understanding helps the learner reformulate facts into useable knowledge. Developing a conceptual framework allows the individual to organize information into meaningful patterns and store it hierarchically in memory. Research on learning provides support for Brandwein's recommendation for major conceptual schemes as the basis for teaching science. Furthermore, contemporary research on learning supports my proposal to use standards as the basis for content, curricular coherence, and greater congruence between scientific inquiry and science teaching.

Responses to Criticism of the Science Curriculum

Criticisms of curricular coherence can be summarized in four themes: *lack of challenging content, lack of instructional focus, inappropriate time to learn,* and *lack of horizontal and vertical connections of content.* Criticisms about the lack of challenging content center on the overemphasis on facts and general lack of a conceptual orientation for science programs. For example, one can ask whether a curriculum is oriented toward scientific concepts that are fundamental to a discipline or topics that may be interesting but do not emphasize scientifically fundamental concepts or processes. *Lack of instructional focus* refers to the lack of depth of treatment of content. For example, content may only receive superficial treatment in the curriculum. *Inappropriate time to learn* refers to the amount of time a concept remains in the curriculum. For example, some concepts are given an inadequate amount of time, while others have a presence much beyond an adequate time for learning to occur. Finally, some concerns about coherence refer to the lack of connections among science concepts and inquiry abilities in both horizontal and vertical dimensions of the curriculum. The cumulative effect of these criticisms is lower student achievement, particularly on national and international assessments, but these qualities can be addressed as issues of curricular design.

To the question of challenging content, school science programs should be based on fundamental or essential scientific concepts and inquiry abilities. Documents such as the *National Science Education Standards* (NRC 1996), *Benchmarks for Science Literacy* (AAAS 1993), *Assessment Frameworks and Specifications 2003* (Mullis et al. 2001), the NAEP 2009 Science Framework (NAGB 2009), and the framework for the OECD-sponsored Program for International Student Assessment (PISA) (OECD 2006) have answered questions about challenging content; each provides a model for what students should know and be able to do. See Figure 2.2 (p. 40) for an example from the national standards.

Figure 2.2

Example of Challenging Content From the *National Science Education Standards*

Challenging content for the science curriculum centers on the conceptual orientation of the standards.

Example from the *National Science Education Standards*: As a result of activities in grades K–4, all students should develop understanding of

• Organisms and Their Environments

However, identifying the content for school science is not enough. One must attend to other curricular and instructional issues. The design of programs must address focus, the depth of treatment for fundamental concepts, and procedures. Figure 2.3 presents an example of focus based on the national standards. Note that the statements include clearly delimited content for the science curriculum and assessments.

Figure 2.3

Example of Instructional Focus From the *National Science Education Standards*

Instructional Focus refers to the depth of treatment of content.
Organisms and Their Environments (Grades K–4)

• All animals depend on plants. Some animals eat plants for food. Other animals eat animals that eat plants.

• An organism's patterns of behavior are related to the nature of that organism's environment, including the kinds and numbers of other organisms present, the availability of food and resources, and the physical characteristics of the environment. When the environment changes, some plants and animals survive and reproduce, and others die or move to new locations.

• All organisms cause changes in the environment where they live. Some of these changes are detrimental to the organism or other organisms, whereas others are beneficial.

• Humans depend on their natural and constructed environments. Humans change environments in ways that can be either beneficial or detrimental for themselves and other organisms.

Source: National Research Council. 1996. *National science education standards*. Washington, DC: National Academies Press.

Appropriate time to learn is closely related to focus. The teaching of science requires one to ask not only what content should be in the curriculum but also the depth to which the content should be developed. Science teachers must judge an appropriate amount of time for students at different developmental stages to learn the content. This is a measure of the opportunities for student learning in lessons, courses, and across the curriculum. I propose that the time to learn some content is quite short; for example, the basic idea that "[a]ll animals depend on plants. Some animals eat plants for food. Other animals eat animals that eat the plants" (NRC 1996, p. 129) can be learned at an introductory level in a relatively short time, perhaps three or four 30-minute lessons. On the other hand, the basic idea that "[a]n organism's patterns of behavior are related to the nature of that organism's environment, including the kinds and numbers of other organisms present, the availability of food and resources, and the physical characteristics of the environment" (NRC 1996, p. 129) may take longer and require a spiraling through different grades with five or six experiences at different grade levels and with exposure for varying amounts of time.

Coherence refers to the number of concepts developed in a uniform set of experiences (for example, lesson, unit, and course) and within a school program (for example, elementary school, middle school, high school, and college). Coherence is a measure of the connectedness among the science concepts that students experience during their study of science. Note that there are both horizontal (that is, across a course) and vertical (that is, between grade levels in school science programs) dimensions to curricula coherence. F. James Rutherford (2000) has written about coherence in high school programs. Rutherford states:

> If coherence in high school science courses is a desirable property, then one can reasonably argue that it should be present at every level of content organization: lessons, units, courses, sequences of courses, and entire curricula. Thus, the topics and activities making up a science lesson or chapter ought to connect with one another to tell a (very limited) story, with, as it were, a discernable beginning, middle, and end. Similarly, the lessons or chapters making up a science unit should connect one another in interesting ways to tell a complete (but still limited) story, and units should connect with one another in interesting ways to tell a more comprehensive story. Notice that two conditions must prevail at each level of organization: All of the parts forming a unit or course must be coherent, and all of those parts must join together to form a conceptual whole. (p. 22–23)

For school science programs, achieving coherence will require curricular designs where less is more—that is, fewer concepts are studied in greater depth. Table 2.2 (p. 42) uses the national standards as an example. In this table, horizontal coherence is modeled within a grade level for a key concept when you read down a column. Vertical coherence between grade level bands is modeled when you move from one column to the next.

Table 2.2

Example of Horizontal and Vertical Coherence From the *National Science Education Standards*

Curricular coherence refers to the connections among concepts in both horizontal and vertical dimensions of the curriculum.

Organisms and Their Environments (Grades K–4)	Populations and Ecosystems (Grades 5–8)	The Interdependence of Organisms (Grades 9–12)
• All animals depend on plants. Some animals eat plants for food. Other animals eat animals that eat the plants. • An organism's patterns of behavior are related to the nature of that organism's environment, including the kinds and numbers of other organisms present, the availability of food and resources, and the physical characteristics of the environment. When the environment changes, some plants and animals survive and reproduce, and others die or move to new locations. • All organisms cause changes in the environment where they live. Some of these changes are detrimental to the organism or other organisms, whereas others are beneficial. • Humans depend on their natural and constructed environments. Humans change environments in ways that can be either beneficial or detrimental for themselves and other organisms.	• A population consists of all individuals of a species that occur together at a given place and time. All populations living together and the physical factors with which they interact compose an ecosystem. • Populations of organisms can be categorized by the function they serve in an ecosystem. Plants and some micro-organisms are producers—they make their own food. All animals, including humans, are consumers, which obtain food by eating other organisms. Decomposers, primarily bacteria and fungi, are consumers that use waste materials and dead organisms for food. Food webs identify the relationships among producers, consumers, and decomposers in an ecosystem. • For ecosystems, the major source of energy is sunlight. Energy entering ecosystems as sunlight is transferred by producers into chemical energy through photosynthesis. That energy then passes from organism to organism in food webs. • The number of organisms an ecosystem can support depends on the resources available and abiotic factors, such as quantity of light and water, range of temperatures, and soil composition. Given adequate biotic and abiotic resources and no disease or predators, populations (including humans) increase at rapid rates. Lack of resources and other factors, such as predation and climate, limit the growth of populations in specific niches in the ecosystem.	• The atoms and molecules on the Earth cycle among the living and nonliving components of the biosphere. • Energy flows through ecosystems in one direction, from photosynthetic organisms to herbivores to carnivores and decomposers. • Organisms both cooperate and compete in ecosystems. The interrelationships and interdependencies of these organisms may generate ecosystems that are stable for hundreds or thousands of years. • Living organisms have the capacity to produce populations of infinite size, but environments and resources are finite. This fundamental tension has profound effects on the interactions between organisms. • Human beings live within the world's ecosystems. Increasingly, humans modify ecosystems as a result of population growth, technology, and consumption. Human destruction of habitats through direct harvesting, pollution, atmospheric changes, and other factors is threatening current global stability, and if not addressed, ecosystems will be irreversibly affected.

Source: National Research Council. 1996. *National science education standards.* Washington, DC: National Academies Press.

For many reasons, the teaching of science has lost coherence (Schmidt et al. 2001). Indeed, the U.S. curriculum has been analyzed compared to top-achieving countries in the Third International Mathematics and Science Study (TIMSS) and was found to lack coherence in ways generally discussed in this chapter.

The examples I used from the national standards demonstrated what might be meant by coherence. Although they served as adequate examples, they were not part of an actual science curriculum. An elementary program developed by BSCS describes an actual curricular framework based on the national standards (see Table 2.3).

Table 2.3

Curriculum Framework for BSCS Science: T.R.A.C.S.

Level	Physical Science	Earth and Space Science	Life Science	Science and Technology
K (Teacher Edition only)	Investigating my world			
1	Investigating properties	Investigating Earth materials	Investigating animals and their needs	Testing materials
2	Investigating position and motion	Investigating weather	Investigating plants	Designing sound systems
3	Investigating electrical systems	Investigating objects in the sky	Investigating life cycles	Designing structures
4	Investigating changing properties	Investigating the changing Earth	Investigating ecosystems	Solving pollution problems
5	Investigating heat and changes in materials	Investigating weather systems	Investigating human systems	Designing environmental solutions

Source: Biological Sciences Curriculum Study (BSCS). 1999. *BSCS Science T.R.A.C.S.* Dubuque, IA: Kendall/Hunt.

The program *BSCS Science: T.R.A.C.S.* serves as the example. As you can see in Table 2.3, the program uses major content themes from the standards and has a coherent vertical and horizontal curricular structure. Although the "grain size" of units differs from that proposed in the 1960s by Brandwein, the focus on major conceptual themes is consistent. I note the contrast of this BSCS program with many contemporary programs often "developed" locally. The local programs usually have a mixture of commercial units, textbook chapters, and assorted activities, all forming a science curriculum. One other salient criticism of many contemporary programs resides in their lack of vertical coherence. Many school districts pay less attention to the organization and development of learning outcomes from grade to grade, compared to learning outcomes within grades. Let me turn to my third theme: congruence.

Congruence and the Teaching of Science

What do I mean by congruence and the teaching of science? *Congruence* refers to a mode of teaching, what the science teacher does, and those tangible and intangible things that contribute to the teaching of science in any particular classroom. The term *congruent* means coinciding, agreeing, or corresponding. In the context of this chapter, congruence parallels Brandwein's use of the term *style*. What, then, would be congruence in the teaching of science?

In the broadest sense, congruence in the teaching of science means that the teacher brings the elements of content and curricular coherence together in ways that optimize learning for students. But this could be said for any discipline. I suggest that congruence in the sciences should center on the theme of scientific inquiry. Teaching science as inquiry brings together content and pedagogy in ways that broaden and deepen student learning and, ultimately, the students' understanding and appreciation of science.

Science as Inquiry?

Although inquiry has been a goal for decades, that does not mean it has been implemented as an integral feature of school programs or science teaching. Indeed, a major synthesis of research published in the 1980s indicated that inquiry did not have a prominent role in science education (Welch et al. 1981). This finding was especially disappointing because there had been discussion of inquiry by prominent individuals (Schwab 1966; Rutherford 1964) and programs that incorporated inquiry as a prominent theme. Like many goals, inquiry provides a rallying point of an apparent common agreement that fosters a sense of community and support among the advocates. Also common to most educational goals, there emerges the need for concrete examples of the abstract ideas and attitudes conveyed by the goal.

In both of his 1960s essays, Paul Brandwein discussed scientific inquiry. In fact, Brandwein devoted considerable space to the various aspects of scientific investigation. Brandwein did not, however, propose inquiry as content. Rather, he assumed it as a strategy or part of the science teachers' style. Just as it is important for students to understand major conceptual themes such as those described by Brandwein, I propose that it is important for students to develop the abilities and understanding of scientific inquiry. I believe that developing an understanding of scientific inquiry is generally consistent with

Brandwein's views. He demonstrated in numerous discussions of science and scientists that he had a clear and deep understanding of inquiry. It seems only reasonable to recommend that any contemporary view of content would include scientific inquiry. Again, the standards provide valuable information for answering questions about scientific inquiry as content and teaching style, to use Brandwein's term.

A New Affirmation of Science as Inquiry

Publication of the *National Science Education Standards* in 1996 gave new life to the inquiry goal. The standards presented inquiry as a prominent theme for both teaching and content. Here is a quotation from that document:

> *Scientific inquiry refers to the diverse ways in which scientists study the natural world and propose explanations based on the evidence derived from their work. Inquiry also refers to the activities of students in which they develop knowledge and understanding of scientific ideas as well as an understanding of how scientists study the natural world.* (NRC 1996, p. 28)

The actual standard states, "As a result of activities in grades 9–12, all students should develop both abilities necessary to do scientific inquiry and understandings about scientific inquiry." Figures 2.4 and 2.5 (p. 46) present the fundamental abilities and understandings associated with these respective components of the standards.

Figure 2.4

Abilities of Scientific Inquiry

- Identify questions and concepts that guide scientific investigations
- Design and conduct scientific investigations
- Use technology and mathematics to improve investigations and communications
- Formulate and revise scientific explanations and models using logic and evidence
- Recognize and analyze alternative explanations and models
- Communicate and defend a scientific argument

Figure 2.5

Understandings About Scientific Inquiry

- Scientists usually inquire about how physical, living, or designed systems function.

- Conceptual principles and knowledge guide scientific inquiries.

- Scientists conduct investigations for a wide variety of reasons.

- Scientists rely on technology to enhance the gathering and manipulation of data.

- Mathematics is essential in scientific inquiry.

- Scientific explanations must adhere to criteria such as the following: A proposed explanation must be logically consistent; it must abide by the rules of evidence; it must be open to questions and possible modification; and it must be based on historical and current scientific knowledge.

- Results of scientific inquiry—new knowledge and methods—emerge from different types of investigations and public communication among scientists. In communicating and defending the results of scientific inquiry, arguments must be logical and demonstrate connections between natural phenomena, investigations, and the historical body of scientific knowledge. In addition, the methods and procedures that scientists used to obtain evidence must be clearly reported to enhance opportunities for further investigation.

The standards shifted the implementation of the teaching-science-as-inquiry theme from an emphasis on "the processes" to cognitive abilities such as reasoning with data, constructing an argument, and making a logically coherent explanation. Furthermore, the standards made it clear that the aim of science education included students' understanding of scientific inquiry. These are elaborated on in *Inquiry and the National Science Education Standards* (NRC 2000), an addenda to the Standards that also includes a summary of research on inquiry.

Concluding Discussion

This chapter used Paul Brandwein's themes of substance, structure, and style as points of departure for a discussion of content, coherence, and congruence as they relate to the teaching of science. For Brandwein, substance referred to major conceptual schemes of the sciences. Structure referred to the curricular organization—that is, how are those conceptual schemes organized and developed. Finally, he referred to style as the complex interactions in the classroom, in particular, what science teachers do to accomplish their goals. In essence, style is the teaching of science. Now, one might ask, "What about the themes of this essay—content, coherence, and congruence?"

The themes establish 21st-century perspectives for Brandwein's themes. In a sense, I used the spirit of science by building on the past, being appropriately skeptical, and applying new ideas that complement and elaborate on Brandwein's original ideas.

Brandwein proposed six conceptual schemes for the science curriculum. These schemes were truly big ideas in the sciences. While I agree with the essential need for conceptual schemes as a central feature of science education, I argue that there is a need for a broader range of content. Specifically, the content described in the original *National Science Education Standards* (NRC 1996) and the common core standards developed early in the 21st century should be the basis for the science curriculum. This recommendation establishes consistent outcomes for science education while leaving open a variety of curriculum options, emphases, and organizations. My recommendations only state that we agree on the content outcomes, not the particular instructional materials that would be implemented to achieve those outcomes. To be clear, I am not recommending a national curriculum. Mine is an argument to agree on ends, and it leaves the means of achieving those ends up to states and local jurisdictions. That is, we agree on a common core of content standards—what students should know and be able to do—not how the curriculum is organized and presented in school science programs

Standards propose that content include not only the physical, life, and Earth-space sciences but also inquiry, technology as it relates to science, personal and social perspectives, and the history and nature of science. We have curricula that are much too crowded with irrelevant information and trivial facts and have too little emphasis on the basic concepts and scientific inquiry. It is well past the time to achieve the proper balance of content in the science curriculum, and the national standards suggest an appropriate and meaningful balance.

Brandwein argued that the substance of science should not be the structure of school science curricula. My point here parallels his, namely that science content is not the science curriculum. In a contemporary perspective, the current curriculum experienced by many students is the metaphorical equivalent to watching television while another person has the channel selector and is clicking through the channels so quickly that the viewer only gets a glimpse of the storylines in dramas, scores in sports, recipes in cooking, sale items in home shopping, headlines in news, or guests on talk shows. Imagine, then, taking a test on the stories, scores, recipes, sales, headlines, and guests. You should get my point about curricular incoherence. Comparing the U.S. science curricula with those of other high-achieving countries demonstrates just how incoherent our programs are. The results of international assessments such as TIMSS and PISA provide evidence about student achievement in the United States as compared to other countries.

In contrast, we should consider a few basic science concepts that have been defined in the standards and use them as the central emphasis for the curriculum. This would help establish curricular coherence in science. I discussed several themes related to curricula coherence, challenging content, appropriate focus, time to learn, and horizontal and vertical connections. My recommendation centers on the need for developing and implementing instructional materials that contribute to greater curricular coherence. As a first step toward meeting this recommendation, I suggest using the standards as the basis for curriculum development. I also noted the existence of some curriculum materials that exemplify curricular coherence, especially when compared to many locally compiled programs.

My third theme was congruence. In the context of this essay, I argue that the teaching of science should be congruent with scientific inquiry. The era represented by Paul F-Brandwein's contributions brought inquiry into science education, especially into the popular lexicon of science teachers and science educators. Most science teachers in the 1960s and 1970s would claim they were "inquiry teachers," especially if they were using programs that have come to symbolize that era; *BSCS Biology, PSSC Physics, CHEM Study,* and *ESCP Earth Science* would be examples. In that era, "teaching science through inquiry" was commonly heard. Inquiry teaching became synonymous with using investigations or doing laboratory activities. Upon close examination, the investigations were designed to facilitate students' learning of content. This was especially true for science teaching at the secondary level.

In addition to implementing activities to enhance conceptual understanding of science, investigations also can be used to develop students' abilities associated with scientific inquiry. The science teacher's goal in this case shifts from an exclusive emphasis on content to facilitating reasoning by asking the student questions about, for example, possible explanations, the role of evidence, alternative explanations, and consistency of current scientific knowledge with students' explanations.

In addition, the teaching of science should include the development of students' understanding of inquiry and the nature of science. These are largely neglected outcomes of science education, yet citizens often encounter situations that require some understanding of science as a way of knowing, as a human endeavor with distinct processes that produce knowledge about the natural world.

Paul Brandwein's works left the science education community with an intellectual investment that had the potential to grow significantly. The 21st-century perspective I have tried to provide here shows how much we can still draw on his work. The teaching of science is even more important today than it was nearly 50 years ago because a sound understanding of science and technology has become essential to our society and the international community. We can take a major step toward improving the teaching of science for all students by systematically and effectively introducing challenging content, increasing curricular coherence, and implementing instructional congruence.

The Science Curriculum and Classroom Instruction

This chapter uses the major contributions of Bob Karplus to develop several important themes that have emerged in the past five decades of curriculum and instruction in science. The chapter begins with a perspective on curriculum development and curriculum developers. A second theme is the importance of identifying major scientific concepts as a foundation for the curriculum. A third theme addresses the importance of incorporating research on learning theory into designs for curriculum and instruction. Finally, there is a theme of curricular reform and the professional development of science teachers. This chapter also presents an occasion for a personal accounting of some intersections of my professional work, in particular at the Biological Sciences Curriculum Study (BSCS), with that of Bob Karplus and his colleagues at the Science Curriculum Improvement Study (SCIS). In the final section, I present some contemporary challenges.

Science Curriculum Development

The science education community owes a great debt to Bob Karplus. One might ask, What is the nature of the debt? And exactly how much is the debt? Briefly, the nature of our debt centers on the characteristics and qualities of curriculum and instruction, the processes of curriculum development, and the place of curriculum developers in science education. As to the debt's amount, I cannot say exactly. After 40 years we are still only paying the interest and have not touched the principal of what we owe.

It should come as no surprise that Karplus developed a theoretical background for science education. He was, by education and inclination, a theoretician. The theoretical background included the nature and development of children's intelligence, the nature and structure of science, and the implications of these two domains for designing science curricula. Karplus was clear about the objective of science education: It was scientific literacy. But his view centered on fundamental conceptual ideas in science and gave less emphasis to a contextual emphasis, such as energy, resources, and environmental quality. He thought that the science curriculum had to provide students with experiences that differed from those they usually had. The unique, unusual, and engaging experience afforded the opportunity for discovery. He proposed the use of experiences with different substances, instruments, environments, and living organisms. Curricular design should be such that students were guided from a point of unique experience to the discovery of a new scientific point of view.

Robert Karplus and, in no small measure, his colleague Herb Thier made this theoretical background into a practical program through SCIS. For the psychological basis of SCIS, they drew on the work of Jean Piaget, Jerome Bruner, and other educators. For the scientific basis, they relied on their own understanding and appealed to Chet Lawson and other colleagues at SCIS.

Principles for the Science Curriculum

In 1969, Karplus prepared an orientation for the Cubberley Curriculum Conference at Stanford University. That presentation was later published as a chapter titled "Some Thoughts on Science Curriculum Development" in *Confronting Curriculum Reform* (Eisner 1971). In this chapter Karplus addressed a view about curriculum specialists and their responsibilities, the process of curriculum development, commercial publication, and curriculum evaluation.

Karplus clearly suggests that he opposed a "teacher proof" view of science curriculum development. Rather, he proposed a variety of objectives and opportunities for attainment. The beginning science teacher could follow the suggested sequence very closely, while an experienced science teacher might adapt the program to accommodate the students' needs and special interests. The view he expressed here is one of shared responsibility for the curriculum.

The chapter Karplus prepared included several principles for science curriculum development. In general, the principles indicated the importance of incorporating different strategies, allowing for spontaneous activities by children, and providing opportunities for social interaction. One principle has insight that seems particularly prescient. I quote:

> *Developmental learning theory is more reliable in the cognitive area; … One implication of this principle is that concept formation should be pursued at low pressure over long periods of time; that is why individual activities have to fit an overall conceptual structure …* (Karplus in Eisner 1971, pp. 58–59)

This quotation contains two wonderful insights about curriculum and instruction—namely, that concept formation is pursued at low pressure over long periods and that activities should have a conceptual structure. Three decades later, a major National Research Council report on learning confirmed these insights. I shall return to this idea in a later section.

His point about commercialization was straightforward. For a program such as SCIS to have a significant effect on schools, it had to be commercially available. This, in turn, meant that the developer had to identify a publisher and arrange for publication.

The chapter also included principles for curriculum evaluation, including that the study has to occur when materials are stable (i.e., after field-testing); it has to last long enough to derive longitudinal results; and it should include all the "stakeholders" in the educational community (the units of evaluation incorporate students, groups of children, teachers, schools, and school districts).

Finally, Karplus defends the existence of curriculum developers. He states that there is value in the specialization of labor here, as in other efforts. For instance, small groups of individuals such as Lawrence Hall of Science (LHS), Biological Sciences Curriculum Study (BSCS), Technical Education Research Center (TERC), and Education Development Center (EDC) can concentrate their time, effort, and resources on the curricular needs of science teachers and the aspirations of the science education community. Through such specialization, science teachers will not be distracted from their primary work by curriculum development. They can concentrate on effective teaching and student learning in their classrooms and schools.

Guidelines for Science Instruction

After a decade of work on SCIS, Karplus had an opportunity to reflect on his work. His reflections resulted in a short statement titled "Three Guidelines for Elementary School Science." In this statement, Karplus first distinguished between the experiential and conceptual aspects of the science curriculum. He indicated the need to consider the relationship between the two, and he summarized a variety of experiences from SCIS, such as electrical and magnetized objects interacting, chameleons eating crickets, and seeds germinating. His point was that students should have direct experiences with physical and biological phenomena. He then went on to make a point that I think provides great insight about these and other examples of experiences one might incorporate in a science curriculum. I quote Karplus at length:

> *Being a physicist, I began my educational activities ten years ago with the notion that force was the fundamental explanatory concept, since force is the cause of motion, and motion is a part of all change. Now I believe that this approach, which is also taken by most physics texts, is not valid. The reason is that observable motion accompanies only a small fraction of phenomena. Thermal, chemical, electrical, optical, and acoustic phenomena do not involve observable motion, hence the force concept is not of direct value in dealing with*

them. Instead, the broader concept of interaction does apply as an explanatory concept for all these areas, and this concept therefore plays the central role in the SCIS program. (Karplus 1971)

In this quotation, Karplus identifies a point that I believe is essential to curriculum and instruction. Students should have deep experiences with fundamental, explanatory concepts. What many students often get in school science is even worse—they have shallow experiences with insignificant topics. This quote from Karplus demonstrates deep educational thinking about scientific understanding, which is different from scientific thinking about educational understanding.

In the second guideline, Karplus argued for using different theories of learning (such as conditioning, discovery, and equilibration) in the construction of curricula. Here, Karplus made explicit the role of learning theories and their appropriate use in curriculum and instruction. One could argue that this guideline expresses common sense more than scientific and educational insight. However, the most cursory review of many school curricula and science instruction will reveal a failure to recognize the assumptions about learning that underlie the program.

In the third guideline, Karplus synthesized the insights from his first and second guidelines, which suggested the need to establish a relationship between students' experiences and science concepts and apply theories of learning and development to instruction. The synthesis became one of the major contributions to science curriculum and instruction—the SCIS learning cycle. The SCIS curriculum developers designed instructional units based on a learning cycle. The original learning cycle consisted of three phases: *exploration*, referring to self-directed investigations; *invention*, referring to the introduction of a new concept; and *discovery*, referring to the application of the same concept in a variety of situations. The following is an extended summary of the SCIS learning cycle.

During exploration, the students gain experience with the environment—they learn through their own actions and reactions in a new situation. In this phase, they explore new materials and new ideas with minimal guidance or expectation of specific accomplishments. The new experience should raise questions or complexities that they cannot resolve with their accustomed patterns of reasoning.

… As a result, mental disequilibrium will occur and the students will be ready for self-regulation. … The second phase, concept introduction, provides social transmission—it starts with the definition of a new concept or principle that helps the students apply a new pattern of reasoning to their experiences. … The concept may be introduced by the teacher, a textbook, a film, or another

medium. This step, which aids self-regulation, should always follow exploration and relate to the exploration activities. ...

In the last phase of the learning cycle, concept application, familiarization takes place as students apply the new concept and/or reasoning pattern to additional situations. (Karplus 1977, pp. 173–174)

Both *A Love of Discovery* (Fuller 2002) and Anton Lawson's book *Science Teaching and the Development of Thinking* (1995) provide an excellent discussion of the origins, applications, variations, and supporting research for the learning cycle. The summaries center on the formative experiences of Karplus, his early classroom activities, working with J. Myron Atkin and publishing "Discovery or Invention?" (Atkin and Karplus 1962), and visiting the Elementary Science Study, where he became familiar with the approach to learning discussed by David Hawkins in "Messing About in Science" (1965). Lawson also discussed the origins of the learning cycle in biology education, which represents an independent and parallel development of the significant features of the learning cycle.

The learning cycle continues to influence curriculum and instruction in science. It has substantial research support (Lawson, Abraham, and Renner 1989) and widespread application through textbooks on science teaching and learning (Lawson 1995; Marek and Cavallo 1997).

New Designs for Curriculum and Instruction

A colleague recently asked what I would do about curriculum and instruction if I returned to the science classroom. I took a moment to reflect, but the response came quickly. I replied, "Concentrate on the instructional core." I would employ an instructional model, concentrate on teaching, and for science content I would implement the best curriculum available. This, however, was not always the case. At a very early point in my career, I would have answered, "Lecture, demonstrations, and occasional laboratories." SCIS played an important role in forming my ideas about curriculum and instruction. I visited SCIS for a week, and that experience helped me form ideas about well-designed curriculum materials and effective instruction. During this period, the work of Jean Piaget became a point of study and eventually a book, *Piaget for Educators* (Bybee and Sund 1982). The latter drew me to the learning cycle and its application in the classroom.

In the late 1960s, I taught SCIS at the University of Northern Colorado Laboratory School. This experience provided more than an academic perspective of the program and the work of Robert Karplus, Herb Thier, and their colleagues. Although this time provided insights that I would use later, I had two especially influential experiences with SCIS. For 30 weeks I taught SCIS, in particular the units on Material Objects and Organisms, to deaf preschool children. In addition, for a brief period I used SCIS with educable mentally retarded children (Bybee 1972; Boekel and Bybee 1973). During this same period, Herb Thier directed a project in which he adapted SCIS materials for the visually impaired (Thier 1971).

I can remember sitting in my office and thinking about experiences that would help the deaf children form ideas about phenomena and how best to sequence those activities. I had to think about conceptual development and not succumb to teaching as telling (which I could do in my ninth-grade Earth science class). The contrast in lessons and teaching between what I could do in my Earth science class and what I had to do in the classes for special students was striking, and the experience had a profound effect on my views of curriculum and instruction. I came to realize the necessity of designing curriculum materials with fundamental concepts and the extraordinary difficulty of teaching for conceptual change.

New Designs for the Science Curriculum

My work in curriculum and instruction was re-engaged in 1985 when I joined the Biological Sciences Curriculum Study (BSCS) and began work as a curriculum developer. IBM supported one of our initial projects—a design study for a program that would eventually be published as *Science for Life and Living: Integrating Science Technology and Health*. The program was revised in the late 1990s and published as *BSCS Science T.R.A.C.S.* The curriculum framework for these programs shows the influence of and variations from SCIS (see Tables 3.1 and 3.2).

Table 3.1

A Curriculum Framework for an Elementary Program

Grade	Science	Technology	Health
1	Objects and properties	Materials and structures	Safety and security
2	Comparison and evidence	Tools and machines	Wellness and personal care
3	Records and data	Construction and testing	Nutrition and dental care
4	Interactions and variables	Problems and solutions	Self and substances
5	Energy chains and food chains	Design and efficiency	Fitness and protection
6	Ecosystems and resources	Constraints and trade-offs	Communication and conflict

Source: Biological Sciences Curriculum Study (BSCS). 1992. *Science for life and living.* Dubuque, IA: Kendall/Hunt.

Table 3.2

A Curriculum Framework for BSCS T.R.A.C.S.

Grade Level	Physical Science	Earth and Space Science	Life Science	Science and Technology
K (Teacher Edition only)	Investigating my world			
1	Investigating properties	Investigating Earth materials	Investigating animals and their needs	Testing materials
2	Investigating position and motion	Investigating weather	Investigating plants	Designing sound systems
3	Investigating electrical systems	Investigating objects in the sky	Investigating life cycles	Designing structures
4	Investigating changing properties	Investigating the changing Earth	Investigating ecosystems	Solving pollution problems
5	Investigating heat and changes in materials	Investigating weather systems	Investigating human systems	Designing environmental solutions

Source: Biological Sciences Curriculum Study (BSCS). 1999. *BSCS Science T.R.A.C.S.* Dubuque, IA: Kendall/Hunt.

New Designs for Science Instruction

As work progressed on the original design study for the *Science for Life and Living*, we realized the need to make some changes in the original SCIS learning cycle. The learning cycle evolved into the BSCS 5E Instructional Model. The 5Es represent engage, explore, explain, elaborate, and evaluate (see Table 3.3, p. 56).

What were the changes we introduced? First, we added an initial phase. We recognized the need for an explicit *engagement* of the student with his or her ideas and those of the curriculum. In its simplest form, this is a question that initiates a student's thinking about an object, organism, or phenomenon. Second, although we maintained the term *exploration* and the original intent of the phase, we recognized the research on cooperative learning (e.g., Johnson, Johnson, and Holubec 1986), so we adapted the exploration phase to include cooperative learning by student groups. We maintained the invention or concept introduction phase but changed the term to *explanation*. For the discovery phase, we again incorporated cooperative learning and changed the phase to *elaboration*. Finally, we added a

Table 3.3

Instructional Model

Stage One
Engage: These activities mentally engage the student with an event or a question. Engagement activities help the students make connections with what they already know and can do.
Stage Two
Explore: The students work with each other, explore ideas together, and acquire a common base of experience, usually through hands-on activities. Under the guidance of the teacher, they clarify their understanding of major concepts and skills.
Stage Three
Explain: The students explain their understanding of the concepts and processes they are learning. The teacher clarifies their understanding and introduces and defines new concepts and skills.
Stage Four
Elaborate: During these activities the students apply what they have learned to new situations, and they build on their understanding of concepts. They use these new experiences to extend their knowledge and skills.
Stage Five
Evaluate: The students assess their own knowledge, skills, and abilities. These activities also focus on outcomes that a teacher can use to evaluate a student's progress.

Source: Biological Sciences Curriculum Study (BSCS) and International Business Machines (IBM). 1989. *New designs for elementary school science and health.* Dubuque, IA: Kendall/Hunt Publishing Company.

phase of *evaluation*. In this phase, students encounter a new activity in which they demonstrate their understanding and abilities.

Between 1985 and 1995, BSCS introduced several new features for curriculum and instruction. We introduced design studies as an initial component of our work. These studies have been published to help other developers and science teachers understand curriculum development. The BSCS design studies include *New Designs for Elementary School Science and Health* (BSCS 1989), *Developing Biological Literacy* (BSCS 1993), and *Making Sense of Integrated Science* (BSCS 2000). These design studies helped pave the way toward new elementary, middle, and high school programs.

The original learning cycle evolved into the BSCS 5E Instructional Model and was adapted to middle and high school programs. During this period, we increasingly recognized the importance and usefulness of the 5E model but also saw the need for professional development that extended beyond an introduction to curriculum materials and instructional models.

In comparison to the late 1950s and early 1960s, in the late 20th and early 21st centuries curriculum design has become much more complex, requiring a higher degree of expertise by developers; thus, implementation has become more complicated, requiring an understanding of science teachers' concerns and extended professional development. To use Karplus's term, communication with teachers has become imperative.

In *Developing Inquiry-Based Science Materials* (2001), Herb Thier and Bennett Daviss report a wonderful story about a Harvard professor who developed an interest in the acoustics of violins. As part of a study he collected a variety of violins, ranging from the cheapest violin to the finest Stradivarius. The professor then erected a small screen in a concert hall and assembled an audience. Yehudi Menuhin, a world-famous violinist, stood behind the screen and played each of the violins for the group. The professor asked the audience to select the best-sounding violin of the collection.

To the professor's amazement, each violin received about the same number of votes. Upon expressing his finding, and shock, to Menuhin, the great violinist provided a significant insight. Menuhin said, "Yes, they sounded about the same. The difference was that the Strad played itself, while I had to work like hell to make the cheap violin sound like anything at all" (Thier and Daviss 2001, p. 6).

I have to wonder what the difference in audience assessment would have been if the violinist would have been average or even a beginner. By analogy, this story tells us something of great science teachers. In addition, it also affords insights about the importance of well-designed curriculum. I hesitate to suggest that curriculum development has reached the equivalent of the finest Stradivarius. However, it is far beyond the cheapest beginning instrument, and certainly is more advanced than something one could design and develop alone, with little support, using parts from other violins, in a short period of time, without formal field testing and evaluation.

Contemporary Challenges for Curriculum and Instruction

Just as earlier generations confronted curriculum reform, now a new generation faces the 21st-century challenges. The 21st century presents the education community with new scientific discoveries, technological advances, national security issues, workforce requirements, and international perspectives. Although these have the power to justify curricular change, it is a paradox that we still must answer fundamental questions. Indeed, the questions Karplus and Their raised in the late 1950s serve us as well today as they did then. How will we answer the following questions?

- How can one create a learning experience that achieves a sure connection between the pupil's intuitive attitudes and the concepts of the modern scientific point of view?
- How can one determine what children have learned?

- How can one communicate with the teacher so that the teacher can, in turn, communicate with the pupils? (Karplus and Thier 1967, p. 11)

With small modifications, these questions set the stage for the following discussion, in which I provide contemporary answers to the questions.

Creating Learning Experiences

The original question has several components: first, students' intuitive attitudes; second, a scientific point of view; and third, learning experiences that connect the first two. I assume that "intuitive attitudes" was Bob Karplus's way of indicating that students have views that are inconsistent with—that is, misconceptions about—scientific concepts. The thrust of this question centers on the application of learning theory to curriculum and instruction. Here, the major report *How People Learn: Brain, Mind, Experience, and School* (Bransford, Brown, and Cocking 1999) and other NRC reports (Duschl, Schweingruber, and Shouse 2007; Michaels, Shouse, and Schweingruber 2008) provide contemporary answers. The findings should inform our decisions about curriculum and instruction.

First, students come to science classrooms with their current conceptions about how the world works. Early in life, children begin making sense of phenomena in their world. In some cases, the ideas children develop provide an accurate foundation for continued learning and development. In some cases, however, the current ideas are misconceptions and hinder learning because the students' current conceptions are very powerful and grounded in concrete experiences, even if they are not correct. If, in contrast, students did not have any ideas about objects, organisms, or phenomena, it would be relatively easy to tell them or give some simple experience that would leave an initial understanding. This, however, is not the case. Achieving sure connections between students' current conceptions and science concepts places demands on the curriculum and classroom instruction. There is a need to draw out students' current conceptions, show their inadequacies, and provide opportunities for students to construct new ideas.

The implications seem clear. Curriculum materials and instructional strategies should incorporate means to identify students' current conceptions and provide time and opportunities for conceptual change. Given the powerful nature of extant concepts (whether consistent or inconsistent with the scientific point of view), the process of conceptual change presents a prominent contemporary challenge to curriculum developers and science teachers.

Second, for pupils to develop competence in an area of inquiry, students must (a) have a deep foundation of factual knowledge, (b) understand facts and ideas in the context of a conceptual framework, and (c) organize knowledge in ways that facilitate retrieval and application. Experts, regardless of their field, draw on a deep and richly structured knowledge base. The ability to plan a task, recognize patterns, formulate reasonable arguments, propose explanations, and analyze alternative explanations is closely related to knowledge. But that knowledge is not a Jeopardy! question or Trivial Pursuit item. That is, it is not disconnected, unrelated, and isolated. The knowledge is connected to and part of a deeper conceptual framework and understanding. This interconnection of factual information and conceptual understanding allows experts to extract meaning from information at levels that novices do not recognize, and it helps experts remember, retrieve, and transfer relevant information. Here, one sees the need for both knowledge and major concepts. The direct implications suggest that curriculum materials be structured using major conceptual ideas and that factual knowledge be introduced in direct and clear connection with the concepts.

Third, students can learn strategies to help them define learning goals for themselves and monitor their progress in achieving these goals. Research on experts who are in the process of inquiry and problem solving reveals that they monitor their own understanding, note where more information is needed, realize the consistency or inconsistency of new information with what they know, and apply analogies that help advance their understanding. Many of these strategies parallel methods of scientific inquiry. Research also has demonstrated that students can be taught strategies such as predicting outcomes, explaining to oneself, activating background knowledge, planning ahead, and noting inadequacies. Research also indicates that these metacognitive strategies must be incorporated into subject matter; they cannot be taught in isolation.

Incorporating inquiry and helping students develop strategies of self-questioning and reflection in the context of their work seems the direct implication of this finding.

So what is the "modern scientific point of view?" It is the content described earlier in the *National Science Education Standards* (NRC 1996) and the contemporary common core standards. The Standards present concepts and abilities appropriate for students at different ages and stages of development. They provide a conceptual framework for curriculum developers and classroom teachers. Table 3.4 (p. 60) summarizes this discussion on creating learning experiences.

In *Why Schools Matter*, Bill Schmidt and his colleagues (2001) state another challenge.

> *When districts provide more demanding and coherent curricula, students find greater opportunities [for learning] and students with greater opportunities usually achieve more. Without national leadership, it would appear that the United States will continue to have only accidental enclaves of excellence.* (Schmidt et al. 2001)

Table 3.4

Creating Learning Experiences

Research on *How Students Learn* (Bransford, Brown, and Cocking 1999, 2000)	Curricular Problem	Instructional Problem	Solution
Students come to classrooms with preconceptions about how the world works.	Identify potential misconceptions	Teach for conceptual change	Instructional model
Competence in an area of inquiry includes (a) a deep foundation of factual knowledge, (b) understanding facts and ideas in the context of a conceptual framework, and (c) organizing knowledge in ways that facilitate retrieval and application.	Incorporate both factual knowledge and a conceptual framework	Establish major concepts, introduce knowledge, and apply knowledge to new situations	Design curriculum based on major scientific concepts and provide experiences to learn factual knowledge and apply it to new situations
Students can learn strategies that help them monitor their progress in problem solving.	Provide opportunities to introduce and emphasize students' metacognitive strategies	Teaching strategies of inquiry	Emphasize inquiry abilities

The National Research Council report *How People Learn* (Bransford, Brown, and Cocking 2000) supports the contention that the curriculum materials used in most classrooms are far from demanding and coherent. Instead, "many models of curriculum design seem to produce knowledge and skills that are disconnected rather than organized into coherent wholes" (Bransford, Brown, and Cocking 2000, p. 138). In the science curriculum in particular, "existing curricula tend to overemphasize facts and underemphasize 'doing science' to explore and test big ideas" (Bransford, Brown, and Cocking 2000, p. 137).

The final challenge in answering the first question requires curriculum and instruction that are rigorous, focused, and coherent. *Rigor* means centering on the content, particularly the conceptual structure of science disciplines. *Focus* refers to the depth of treatment of the content. *Coherence* refers to the connections among science concepts and inquiry abilities in both horizontal and vertical dimensions of the curriculum.

Knowing What Students Have Learned

Eventually, educators get to the assessment question. To provide a contemporary answer, I appeal to several resources. The first is another National Research Council report, *Knowing What Students Know* (Pellegrino, Chudowsky, and Glaser 2001), the second is *Understanding by Design* (Wiggins and McTighe 2005), and the third is *Classroom Assessment and the National Science Education Standards* (Atkin, Black, and Coffey 2001). The first book provides a theoretical perspective, one that parallels *How People Learn*, and the latter two help science teachers with the practical problems of incorporating assessment into curriculum and instruction.

The book *Knowing What Students Know* develops the idea that effective assessment uses three interrelated elements—a model of student cognition, a set of beliefs about the observation that will provide evidence of learning, and the interpretation of that evidence. All three elements must be coordinated and synchronized. In an example, a teacher listens, asks a true/false or multiple-choice question, then interprets the student's response to the question. In most instances, the question mirrors a statement from the book or lecture, and the evidence suggests the student did or did not return the information. In this example, the science teacher did not make explicit a model of learning (it was in fact the blank slate model) and likely did not review the assessment prior to formulating the test questions. Science teachers can do better than this.

Understanding by Design describes a process that will bring science teachers closer to determining what students have learned. The process is called *backward design* (Figure 3.1, p. 62). Conceptually, the process is simple. Begin by identifying your desired learning outcomes, for example, concepts, knowledge, or abilities. Then determine what would count as acceptable evidence of student learning and design an assessment that will provide the acceptable evidence. Then, and only then, begin developing the activities that will provide students with opportunities to learn.

The BSCS 5E Instructional Model and the *National Science Education Standards* clarify the process. Let us say you identified the desired learning as "Life Cycles of Organisms." One would review concepts and determine the acceptable evidence of learning. For instance students would need to be able to identify life cycles of plants and animals and describe aspects of the cycle (e.g., being born, growing to adulthood, reproducing, and dying). You might expect students to recognize that offspring closely resemble their parents and that some characteristics are inherited from parents while others result from interactions with the environment. One could design an *evaluate* activity, such as growing Fast Plants under different environmental conditions, then design a rubric with the aforementioned criteria. Then one would proceed to design the *engage, explore, explain,* and *elaborate* experiences. If necessary, the process would be iterative between the evaluate and other activities as the development process progresses.

Figure 3.1

Backward Design Combined With the BSCS 5E Instructional Model

1   Identify desired results *(National Standards)*

2   Determine acceptable evidence of learning
Design *evaluate* activities

3   Develop learning experiences and activities
Engage, explore, explain, elaborate

Source: Wiggins, G., and J. McTighe. 2005. *Understanding by design.* Expanded 2nd ed. Alexandria, VA: Association for Supervision and Curriculum Development (ASCD).

Communicating With Science Teachers

I like the way Karplus formulated this question as a communication problem. I would extend this a bit further by suggesting that the answer lies in the realm of professional development. Here I appeal to a book by Susan Loucks-Horsley and her colleagues, *Designing Professional Development for Teachers of Science and Mathematics* (2003).

Upon reflection, communicating new knowledge and developing new skills for science teachers may present the most difficult set of problems and the most critical barriers to effective curriculum reform. We can begin by returning to *How People Learn* and the assumptions that science teachers have about curriculum and instruction. Adapted statements from *How People Learn: Bridging Research and Practice* (Donovan, Bransford, and Pellegrino 1999, pp. 1–13) will help inform the discussion.

- Science teachers come to educational reform with ideas about curriculum and instruction. If their understanding is not engaged, new concepts and information may not be communicated, may be interpreted in light of current practices, or may be rejected.
- To develop competency in the area of curriculum and instruction, science teachers should (a) have a deep foundation of knowledge about the science discipline and student learning, (b) understand these concepts and ideas in the context of frameworks for curriculum and instruction, and (c) organize knowledge in ways that facilitate application in science classrooms.

- Science teachers can learn strategies that help improve curriculum and instruction.

I propose that professional development (i.e., communication) might take the same form as the responses to similar problems for student learning. First, it may be useful to use an instructional model for the introduction of a new curriculum. Second, we need to provide experiences and opportunities for science teachers to develop a foundation of knowledge. At this stage, some attention should be directed to the frameworks for curriculum and instruction and applications for the classroom.

Finally, a deep and thorough analysis of instructional materials provides professional development experiences that help science teachers develop insights about curriculum and instruction and acquire strategies that will improve their understanding and use of curriculum materials and instructional models (see Table 3.5, p. 64).

Concluding Discussion

Robert Karplus joined a generation of scientists who provided leadership in curriculum reform. They changed the future of curriculum and instruction. Indeed, the ideas of Robert Karplus, perhaps more than anyone else, presented a fundamental shift in the design of instructional materials for elementary school science. Other contemporaries, such as Robert Gagne, applied learning theory to the design of *Science: A Process Approach*, and David Hawkins did so in *Elementary Science Study*. The ideas in SCIS have persisted for more than 40 years and influenced other programs such as elementary, middle, and high school programs at BSCS.

The shift in curriculum and instruction was fundamental. First, curriculum development became a specialized work of groups such as Lawrence Hall of Science (LHS), the Educational Development Center (EDC), and the Biological Sciences Curriculum Study (BSCS). Science teachers no longer had to assume the task of both developing curriculum materials and teaching. They could select the best materials and adapt those materials to their students' unique needs. Second, the implementation of curriculum materials became a central means of educational reform. The educational community realized the central role of instructional materials as a means of education reform. This, to my way of thinking, keeps reform close to the instructional core and the needs of science teachers, and at the heart of the teaching-learning process. Third, learning theories became the basis for more systematic instruction. These theories were applied through instructional models that were integral to the curriculum.

Let me characterize this change in our views of curriculum and instruction. In 1903, Wilbur and Orville Wright made the first powered flight in a plane of

Table 3.5

Communicating About Curriculum and Instruction

Assumption	Goal	Process	Activity
Science teachers come to education reform with ideas about curriculum and instruction. If their understanding is not engaged, new concepts and information may not be communicated, may be interpreted in light of current practices, or may be rejected.	Identify current ideas about curriculum and instruction	Introduce alternative ideas about curriculum and instruction	Follow instructional model
To develop competency in the area of curriculum and instruction, science teachers should (a) have a deep foundation of knowledge about science disciplines and student learning, (b) understand these concepts and ideas in the context of frameworks for curriculum and instruction, and (c) organize knowledge in ways that facilitate application in science classrooms.	Clarify the relationship between student experiences and science concepts	Establish major ideas and knowledge through the introduction of new curricular materials	Analyze instructional materials
Science teachers can learn strategies that help improve curriculum and instruction.	Provide opportunities and investigate instructional materials	Develop the capacity to analyze instructional material	Implement new curriculum materials

their own invention and construction. Now, a hundred years later, airplanes are highly sophisticated technological systems, and the pilots do not invent and construct their own planes. Many school districts and science teachers harbor the idea that they can invent and construct a science curriculum. The time has long passed when this represented a reasonable approach to effective curriculum or educational reform. The task of curriculum development requires the work of professionals who specialize in this work.

Robert Karplus likely did not realize the long-term consequence of his career in science education. Many of his explorations and inventions anticipated discoveries more than 40 years later. His insight and intellect resulted in advances in curriculum and instruction that carry the science education community well into the 21st century.

The contemporary goals such as leaving no child behind or racing to the top involve implementing challenging curriculum materials and developing high-quality teachers for every classroom. We may have to look to the past to see the future.

Teaching Science as Inquiry

The goal of teaching science as inquiry has a long history in American education. Although terms used to describe this aim have varied, the goal has been a priority since the 1800s. The way inquiry has been interpreted by policy makers, included in programs, and implemented by teachers has an equally long and varied history. The emphasis on inquiry as a goal of science education programs has included teaching strategies and learning outcomes, the former being the dominant interpretation by policy makers, curriculum developers, and classroom teachers.

Now, in the early years of the 21st century, the science education community must support science teachers as they respond to contemporary scientific, social, and economic challenges that require a new emphasis on teaching science as inquiry.

A Brief History to 1957

In the United States, science itself had little recognition prior to the mid-19th century: "[F]aith was at least as important as empirical data and in many instances it dominated the practices of science. This faith was often a complex mixture of Christian theology, idealism, and entrenched traditions" (Stedman 1987, p. 657). So it is no surprise that discussions of science education, including scientific inquiry and laboratory work, were absent as well (Bybee and DeBoer 1993; DeBoer 1991).

The public's interest in science and the scientific method increased in the late 19th century. Most likely because of scientific progress in physics, chemistry, and biology and technological advances associated with the industrial revolution, some Americans proposed that scientific thinking was needed to help the public address everyday problems. As reported by historian John Rudolph

(2005), "One eminent scientist in 1884 argued for a thorough reorganization of higher education around the teaching of the scientific method" (p. 346).

During this period, several individuals brought science into discussions of school and college curricula. Louis Agassiz at Harvard University provided an early example of teaching science as inquiry "when he invited students to visit his lab, study specimens firsthand, and thereby gain direct knowledge. He directed field trips to the countryside and seashore, encouraged students to make their own collections, and conducted instruction by correspondence with specimen collectors around the country" (Stedman 1987, p. 660).

The Harvard List of Experiments

Charles W. Eliot, a chemist and president of Harvard, articulated the need for science and established the laboratory as an essential part of science instruction in American high schools (Stedman 1987). Eliot asked the physics department at Harvard to develop an entrance requirement that emphasized the laboratory as part of high school physics courses. The prestige of Harvard all but assured the list of experiments would become a part of high school science programs. By 1889, the list was published as "Harvard University Descriptive List of Elementary Physical Experiments" and covered a wide range of physics topics.

The "Harvard List of Experiments" became more than a laboratory notebook and entrance requirement. It first became the basis for a physics course and later for a national course in physics that was part of the newly formed College Entrance Examination Board's requirement. The use of this descriptive list of experiments and influences of scientists was furthered by the National Education Association's Committee of Ten report. The widespread acceptance of this report became the defacto first voluntary national standards for science, and the roles of laboratory experiences and inquiry as teaching methods were clearly part of the standards.

The Committee of Ten

In the United States, the laboratory method moved from broad goal statements, particularly for high school education, to the recommendations of policy in the 1893 NEA "Report of the Committee of Ten on Secondary School Studies." The report underscored the importance of science for all students, whether they intended to go to college or enter the workforce. Of significance to this chapter is that the report underscored the "absolute necessity of laboratory work" (NEA 1894, p. 27).

The prominent role of science in the Committee of Ten report undoubtedly was influenced generally by the industrial revolution and specifically by two scientists, both college presidents, Charles W. Eliot (Harvard) and Ira Remsen (Johns Hopkins).

The degree to which the laboratory became a part of high school science programs no doubt varied. Those schools that fed prominent universities tried to meet the standards; those without resources maintained the historic and cost-effective lecture-recitation approach. The significant increase in student enrollment beginning in the late 1800s (circa 1886–1900) contributed to the reluctance of school administrators and science teachers to embrace the laboratory approach to science. This lack of support for the laboratory, in particular the Harvard list of experiments, was aided by scientists such as C. R. Mann and organizations such as the Central Association for Science Mathematics Teaching (CASMT). The plea was for greater personal and social relevance of physics by revising the Harvard list to include greater emphasis on qualitative laboratory work (Rudolph 2005).

This shift represented the emergence of two perspectives on the goals of science education in general and the role of the laboratory in particular. These ideologies are evident and still in conflict in contemporary forms. The conflict and apparent opposition is between utility and inquiry. The apparent view that these goals are incompatible continues to this day.

The Influence of John Dewey

In 1910, John Dewey published a small book titled *How We Think*. In this book, Dewey introduced what he called a *complete act of thought*. According to Dewey, a complete act of thought consisted of five logically distinct steps: (i) a felt difficulty; (ii) its location and definition; (iii) suggestions of possible solutions; (iv) development by reasoning of the bearings of the suggestion; and (v) further observation and experiment, leading to its acceptance or rejection—that is, the conclusion of belief or disbelief (Dewey 2005, p. 60).

There are several reasons for mentioning Dewey's book and logical phases based on his conception of a complete act of thought. First, the book title, *How People Learn* (Bransford, Brown, and Cocking 1999), anticipates a contemporary synthesis of research on learning. Second, the steps Dewey described established what became the five steps of the scientific method that has influenced science teachers' conception of scientific inquiry. Finally, the five phases also anticipate the role of instructional models such as the BSCS 5Es.

The fact that Dewey's five phases became a rigid sequence introduced in science textbooks and classrooms is unfortunate. John Dewey did not perceive the methods of science as a lockstep process. Just the year before publishing *How We Think*, Dewey addressed the American Association for the Advancement of Science meeting on the topic "Science as Subject-Matter and As Method" (Dewey 1910). In his address and published article, he argued for the importance of using the scientific method in school science programs and presented a dynamic view of inquiry.

Early in the 1910 publication of Dewey's address, he states his position by saying, "I mean that science has been taught too much as an accumulation of ready-made material with which students are to be made familiar, not enough as a method of thinking, an attitude of mind, after a pattern of which mental habits are to be transformed" (Dewey 1910, p. 121). Dewey elaborated on scientific method as a habit of mind. One should also notice that in the following excerpts, Dewey refers to aims that include the abilities of inquiry, the nature of science, and an understanding of subject matter.

Surely if there is any knowledge which is of most worth it is knowledge of the ways by which anything is entitled to be called knowledge instead of being mere opinion or guess work or dogma.

Such knowledge never can be learned by itself; it is not information, but a mode of intelligent practice, and habitual disposition of mind. Only by taking a hand in the making of knowledge, by transferring guess and opinion into belief authorized by inquiry, does one ever get a knowledge of the method of knowing. (p. 125)

But that the great majority of those who leave school have some idea of the kind of evidence required to substantiate given types of belief does not seem unreasonable. Nor is it absurd to expect that they should go forth with a lively interest in the ways in which knowledge is improved by a marked distaste for all conclusions reached in disharmony with the methods of scientific inquiry. (p. 127)

Later Dewey again states his position. "Thus we again come to the primary contention of the paper: That science teaching has suffered because a science has been so frequently presented just as so much ready-made knowledge, so much subject-matter of fact and law, rather than as the effective method of inquiry into any subject matter" (p. 127).

In a later section of his address, Dewey makes his position clear for the third time. The perspective expressed by Dewey in 1909 is even applicable now, more than 100 years later.

I do not mean that our schools should be expected to send forth their students equipped as judges of truth and falsity in specialized scientific matters. But that the great majority of those who leave school should have some idea of the kind of evidence required to substantiate given types of belief does not seem unreasonable. Not is it absurd to expect that they should go forth with a lively interest in the ways in which knowledge is improved and a marked distaste for all conclusions reached in disharmony with the methods of scientific inquiry. (Dewey 1910, p. 127)

Dewey concludes with this powerful statement:

One of the only two articles that remain in my creed of life is that the future of our civilization depends upon the widening spread and deepening hold of the scientific habit of mind; and that the problem of problems in our education is therefore to discover how to mature and make effective this scientific habit. (p. 127)

I have quoted John Dewey at length because 100 years ago he articulated the need for teaching science as inquiry, for which he included several important outcomes: developing thinking and reasoning, formulating habits of mind, learning science subject matter, and understanding the processes of science. Dewey later wrote *Logic: The Theory of Inquiry* (1938), in which he presented his "steps" in the scientific method (induction, deduction, mathematical logic, and empiricism). This book no doubt influenced the many science textbooks that treat the scientific method as a fixed sequence as opposed to a variety of strategies whose use depends on the question being investigated and on the researchers. Discussions about the role of the scientific method in science classrooms and textbooks continue in the science education community. I think it is clear that John Dewey did not support teaching the scientific method as a formal step-by-step sequence. He likely did support phases of instruction based on the psychology of learning.

The historian John Rudolph (2005) has proposed that educators quickly embraced the five steps for the following reasons: (1) the steps' alignment with the trends toward the psychology of students as applied in problems from actual life situations, (2) increasing levels of enrollment in schools, and (3) the ease of applying scientific approaches without attending to the nuances of individual and contextual differences. In the end, a complex set of social, educational, and scientific trends led educators to equate Dewey's idea of reflective thought with the scientific method. Soon the scientific method was included in textbooks, thus becoming part of the knowledge that students had to memorize.

The Harvard Red Book

In 1945, a Harvard committee published *General Education in a Free Society*. The report included a section on science and mathematics in the secondary schools. After a fairly extensive discussion of what science is, what science is not, what scientists do, and the ways scientists adapt the modes of inquiry, the committee summarized their view of scientific inquiry:

The working scientist brings to bear upon these problems everything at his command—previous knowledge, intuition, trial and error, imagination, formal logic, and mathematics—and these may appear in almost any order in the course of working through a problem. … The nub of the matter is that the problem be solved. (Harvard Committee 1945, p. 158)

From a historical point of view, these criteria by a prestigious committee came shortly after the education community embraced the scientific method and *solidified* its place in science education by placing the five steps in textbooks, often in the first chapters of science textbooks.

The Influence of James B. Conant

James B. Conant was president at Harvard and by nature of his position had appointed the committee and attended to the report's conclusions. Conant had a particular interest in science because he was a chemist. The view expressed later by Conant in *Science and Common Sense* (1951) should not come as a surprise. I refer to Chapter 3, titled "Concerning the Alleged Scientific Method." After an introduction, Conant states his view in no uncertain terms.

There is no such thing as the *scientific method. If there were, surely an examination of the history of physics, chemistry, and biology would reveal it … few would deny it is the progress in physics, chemistry, and experimental biology which give everyone confidence in the procedures of the scientist. Yet a careful examination of those subjects fails to reveal any* one *method by means of which the matters in these fields broke new ground.* (Conant 1951, p. 45; emphasis in original)

Expressions of the scientific method have continued, especially in opening chapters of science textbooks (Lederman 1992). As we enter the 21st century, it is time for science teachers to introduce accurate and appropriate perspectives of scientific inquiry.

The Recent Past, 1957 to Present

I selected 1957—in fact, October 4, 1957—as the place to begin this discussion because the science education community continually refers to Sputnik as initiating a major era of reform. Although Sputnik-era reform implemented any number of innovations on science education, teaching science as inquiry would have to be among those that have been sustained for more than 50 years since that reform.

The Influence of Joseph Schwab

One of the intellectual leaders of the Sputnik reform was Joseph Schwab, whose extensive writing established teaching science as inquiry as a prominent theme for the era. In 1960, Schwab published "Enquiry, the Science Teacher, and the Educator" (*enquiry* was Schwab's preferred spelling of the term) in *The Science*

Teacher. In the following discussion, the reader should note the parallel between the concepts of normal and revolutionary science as described by Thomas Kuhn in *The Structure of Scientific Revolutions* (Kuhn 1970) and Schwab's use of stable and fluid enquiry.

In this article for science teachers, Schwab presented a distinction between what he called stable scientific enquiry and fluid enquiry. Stable enquiry is the pursuit of scientific investigations centered on answering questions that fill in knowledge at places where there is incomplete understanding of particular scientific principles. Those principles are the origins and guiding ideas of enquiry, and the result is greater understanding of that particular principle.

Fluid enquiries have the intention of testing, revising, or ultimately inventing new principles. According to Schwab, both approaches have value as stable enquiry completes and fills in knowledge of a particular principle and fluid enquiry invents new principles. Both types of enquiry advance scientific knowledge, but Schwab argued that fluid enquiry was essential to the long-term advancement of scientific knowledge.

Closely related to our contemporary situation, Schwab stated that England, France, Germany, and Scandinavia trained their potential scientists with an emphasis on fluid inquiry. Schwab brings his views on enquiry to education when he states the following:

> *We are asked to discover, select, motivate, and launch an increasingly large group of fluid enquirers and original engineers—and a non science public which understands the nature and consequences of the work these scientists do.* (Schwab 1960, p. 8)

In the late 1950s and 1960s, Joseph Schwab published other articles on inquiry. Schwab laid the foundation for the emergence of inquiry as a prominent theme in the curriculum reform of that era (Schwab 1958, 1960, 1966). Schwab grounded his argument to teach science as inquiry in science itself: "The formal reason for a change in present methods of teaching the sciences lies in the fact that science itself has changed. A new view concerning the nature of scientific inquiry now controls research" (1958, p. 374).

When Schwab discussed the implication of these changes for education, he quickly pointed out that science textbooks and science teachers were presenting science in a way that was inconsistent with modern science. According to Schwab (1966, p. 24), science was taught " … as a nearly unmitigated *rhetoric of conclusions* in which the current and temporary constructions of scientific knowledge are conveyed as empirical, literal, and irrevocable truths." Schwab goes on to clarify his assertion: "A rhetoric of conclusions, then, is a structure of discourse which persuades men to accept the tentative as certain, the doubtful as the undoubted, by making no mention of reasons or evidence for what it asserts, as if to say, 'this, everyone of importance knows to be true'" (1966, p. 24).

The implications of Schwab's ideas were, for their time, profound. He suggested first that science should be presented as inquiry, and second that students should undertake inquiries as the means to learn science. To achieve these changes, Schwab (1960) recommended that science teachers first look to the laboratory and use these experiences to lead rather than lag the classroom phase of science teaching. That is, the laboratory experience should precede rather than follow the formal explanation of scientific concepts and principles. He also suggested that science teachers consider three levels of openness in their laboratories. First, the materials can be used to pose questions and describe methods to investigate the questions that allow students to discover relationships they do not already know. Second, the laboratory manual or textbook can pose questions, but the methods and answers are left open. Finally, in the most open approach, students confront phenomena without textbook- or laboratory-based questions. They are left to ask questions, gather evidence, and propose explanations based on their evidence.

Schwab also proposed a second approach, which he referred to as *enquiry into enquiry*. In this approach, teachers provide students with readings, reports, or books about research. They engage in discussions about the problems, data, role of technology, interpretation of data, and conclusions reached by scientists. Where possible, students should read about alternative explanations, experiments, debates about assumptions, use of evidence, and other issues of scientific inquiry.

Joseph Schwab had a tremendous influence on the original design of instructional materials—the laboratories and invitations of inquiry—for the Biological Sciences Curriculum Study (BSCS). Schwab's recommendation paid off in the late 1970s and early 1980s when education researchers asked questions about the effectiveness of these programs. Shymansky (1984) reported evidence supporting his conclusion that "BSCS biology is the most successful of the new high school science curricula" (p. 57).

Curriculum reform was a centerpiece of the Sputnik era, and Joseph Schwab was among the intellectual leaders supporting the idea of teaching science as inquiry as a fundamental part of the reform. The founding of BSCS in 1958 brought together Joseph Schwab and curriculum reform in a fruitful union. Indeed, inquiry has been recognized as one of the significant education legacies of the BSCS project (Rudolph 2008). The period leading up to the founding of BSCS had witnessed the Cold War and a "red scare," with the subsequent threat to intellectual freedom. This atmosphere contributed to the linking of scientific inquiry and intellectual freedom as scientists appealed to policy makers, members of the business and industry communities, and citizens to maintain the integrity of science. As BSCS took form, the founders and initial staff and steering committee—including Bentley Glass, Arnold Grobman, H. J. Muller, Paul Brandwein, John A. Moore, and Joseph Schwab established scientific inquiry as a central learning outcome and guiding principle for future curriculum materials.

Did the Science Education Community Meet the Challenge of Teaching Science as Inquiry?

When this question centers on teaching science as inquiry, the answer has to be no. As early as the mid-1960s, insightful criticism emerged. For example, in 1964, F. James Rutherford addressed the role of inquiry in science teaching. He began the article by pointing out that when it comes to the teaching of science, we are unalterably opposed to rote memorization, and we are all for the teaching of scientific processes, critical thinking, and the inquiry method. Rutherford also noted that the practice of science teaching does not represent science as inquiry; in fact, the idea of "teaching science as inquiry" needs clarification. He shows how the terms are sometimes used in a way that emphasizes that inquiry is really part of the science content itself. Science teaching can help students learn about inquiry. At other times, educators refer to a particular technique or strategy for teaching science content. That is, students can conduct an inquiry to learn science concepts and principles.

Rutherford presented these observations:

1. It is possible to gain a worthwhile understanding of science as inquiry once we recognize the necessity of considering inquiry as content and operate on the premise that the concepts of science are properly understood only in the context of how they were arrived at and what further inquiry they initiated.

2. As a corollary, it follows that it is possible to learn something of science as inquiry without the learning process itself having to follow precisely any one of the methods of inquiry used in science. That is, inquiry as technique is not absolutely necessary to understanding inquiry as content.

3. Although the laboratory can be used to provide the student experience with and knowledge of some aspects or components of the investigative techniques employed in a given science, it can effectively do so only after content of the experiments has been carefully analyzed for usefulness in this regard.

Rutherford connected teaching science as inquiry and the knowledge base for doing so. He concluded that until science teachers acquire "a rather thorough grounding in the history and philosophy of the sciences they teach, this kind of understanding will elude them, in which event not much progress toward the teaching of science as inquiry can be expected" (Rutherford 1964, pp. 80–84).

In the late 1970s and early 1980s, the National Science Foundation (NSF) supported a project that synthesized a number of national surveys, assessments, and case studies about the status of science education in the United States (Harms and Kohl 1980; Harms and Yager 1981). One major portion of this review centered on the role of inquiry in science teaching and was completed by Wayne Welch, Leo Klopfer, Glen Aikenhead, and James Robinson (1981). Their analysis revealed that the science education community used the term *inquiry* in a variety of ways, including the general categories identified in this review—inquiry as content and inquiry as instructional technique—and generally science educators and science teachers were unclear about the term's meaning. The researchers identified several discrepancies that presented doubts about the implementation of inquiry in either use of the term. The greatest discrepancy was between teachers' espoused belief in the importance of teaching science as inquiry and their actual practice. The evidence indicated that "although teachers made positive statements about the value of inquiry, they often felt more responsible for teaching facts, 'things which show up on tests,' 'basics' and structure and the work ethic" (Welch et al. 1981).

Teachers expressed a number of reasons for not teaching science as inquiry, introducing the content (knowledge and abilities), or using inquiry-oriented experiences. Among the reasons cited were problems with classroom management, difficulty meeting state requirements and obtaining supplies and equipment, dangers for students, and concerns about whether inquiry really worked. Notice that the justification centers on inquiry as instructional technique. In conclusion, the authors (Welch et al. 1981) reported,

> The widespread espoused support of inquiry is more simulated than real in practice. The greatest set of barriers to the teacher support of inquiry seems to be its perceived difficulty. There is legitimate confusion over the meaning of inquiry in the classroom. There is concern over discipline. There is worry about adequately preparing children for the next level of education. There are problems associated with the teachers' allegiance to teaching facts and to following the role models of the college professors. (p. 40)

I participated on the analysis of biology for Project Synthesis, and that team concluded, "In short, little evidence exists that inquiry is being used" (Hurd et al. 1980, p. 391).

In 1986, Kenneth Costenson and Anton Lawson pursued answers about the lack of inquiry by surveying a group of biology teachers. Teachers gave the following reasons for not teaching biology as inquiry: (1) lack of time and energy (e.g., it takes too much time to develop inquiry materials), (2) too slow (e.g., using inquiry takes too much time, so the district curriculum will not be covered) , (3) reading too difficult (e.g., students cannot read the inquiry book), (4) risk too high (e.g., administration will be critical of teaching), (5) tracking (e.g., level of thinking is too high for students in regular biology), (6) student immaturity (e.g., students waste too much time in inquiry experiences), (7) teaching habits (e.g., I cannot change my style of teaching), (8) sequential material (e.g., I cannot skip chapters and labs in inquiry textbooks), (9) discomfort (e.g., inquiry teaching makes me feel uncomfortable, not in control), and (10) too expensive (e.g., it will cost too much to equip the lab for inquiry) (Costenson and Lawson 1986, p. 151). Their survey responses were similar to those reported by Welch et al. in 1981. Although the context for the Costenson and Lawson study was biology, similar results would likely be obtained for other disciplines, particularly at the secondary level. I list all 10 reasons because they form the substantial barriers between policies—for example, the *National Science Education Standards* (NRC 1996) that recommend science as inquiry and science programs that incorporate teaching science as inquiry and the actual practices in science classrooms.

Costenson and Lawson (1986) conclude their article by saying,

In our opinion, all ten of the previous reasons for not using inquiry are not sufficient to prevent its use. However, to implement inquiry in the classroom we see three crucial ingredients: (1) teachers must understand precisely what scientific inquiry is; (2) they must have sufficient understanding of the structure of biology itself, and (3) they must become skilled in inquiry teaching techniques. (p. 158)

In this quotation, we again see the differentiation of inquiry as *content* to be understood first by teachers and then by students and inquiry as a *technique* to be used by teachers to help students learn biology.

Project 2061

In 1985, F. James Rutherford inaugurated Project 2061, a long-term initiative of the American Association for the Advancement of Science (AAAS) to reform K–12 education. Project 2061 materials such as *Science for All Americans* (AAAS 1989) and *Benchmarks for Science Literacy* (AAAS 1993) have made significant statements about teaching science as inquiry.

In *Science for All Americans* (AAAS 1989), the lead chapter discusses the nature of science, and another chapter discusses "Historical Perspective." These chapters provide the basis for recommendations for including scientific inquiry in school programs. Rutherford and Project 2061 made concrete recommendations consistent with his 1964 critique. The chapter "Habits of Mind" includes categories of values and attitudes, manipulation and observation, communication, and, very important, critical-response skills.

In a separate chapter, "Effective Learning and Teaching," *Science for All Americans* (AAAS 1989, pp. 147–149) has a general recommendation: "Teaching Should Be Consistent With the Nature of Scientific Inquiry," followed by specific recommendations:

- Start with questions about nature
- Engage students actively
- Concentrate on the collection and use of evidence
- Provide historical perspectives
- Insist on clear expression
- Use a team approach
- Do not separate knowing from finding out
- Deemphasize the memorization of technical vocabulary

Benchmarks for Science Literacy (AAAS 1993) provides actual learning outcomes for the aforementioned chapters on the nature of science, historical perspectives, and habits of mind. In addition, there is an excellent research base that indicates what students should know and be able to do relative to various benchmarks. Project 2061 also set in place goals and specific benchmarks for the teaching aspect of scientific inquiry as content and made recommendations for using teaching techniques associated with inquiry. The work of this project clearly set the stage and influenced the *National Science Education Standards* (NRC 1996).

National Science Education Standards

More than a decade ago, the *National Science Education Standards* (NRC 1996) presented national policies that included teaching science as inquiry. Release of the standards again brought the issue of teaching science as inquiry to the forefront in the education community. In the *National Science Education Standards* (NRC 1996), scientific inquiry refers to several related but different aspects of teaching and learning: the ways scientists study the natural world, activities of students, strategies of teaching, and outcomes that students should learn. The *National Science Education Standards* provides the following statement on scientific inquiry:

> *[I]nquiry is a multifaceted activity that involves making observations; posing questions; examining books and other sources of information to see what is already known; planning investigations; reviewing what is already known in light of experimental evidence; using tools to gather, analyze, and interpret data; proposing the results. Inquiry requires identification of assumptions, use of critical and logical thinking, and consideration of alternative explanations.* (NRC 1996, p. 23)

The Standards use the term *inquiry* in two ways. First, inquiry is *content*, which is divided between what students should *understand* about scientific inquiry and the *abilities* students should develop based on their experiences with scientific inquiry. Second, the term *inquiry* refers to teaching strategies and the processes of learning associated with inquiry-oriented activities. In this section, I address the content, beginning with the following statement of the content standard for Science as Inquiry for grades 9–12 (see Figure 4.1).

Figure 4.1

Content Standard for Science as Inquiry

General Standards for Inquiry
All students should develop

- abilities necessary to do scientific inquiry.
- understandings about scientific inquiry.

Science as Inquiry: The Abilities

Figure 4.2 presents the key abilities from the standard. Based on the original discussion in the Standards, this discussion provides details about the fundamental abilities. As you read the descriptions, note the distinct emphasis on *cognitive* abilities and critical thinking by students. This emphasis differentiates the Standards from the traditional emphasis on processes without eliminating activities such as students' observing, inferring, and hypothesizing. In this sense, the Standards advance our understanding of inquiry beyond processes (Millar and Driver 1987).

Figure 4.2

Science as Inquiry: Fundamental Abilities for Grades 9–12

Fundamental Abilities Necessary to Do Scientific Inquiry

- Identify questions and concepts that guide scientific investigations.
- Design and conduct scientific investigations.
- Use technology and mathematics to improve investigations and communications.
- Formulate and revise scientific explanations and models using logic and evidence.
- Recognize and analyze alternative explanations and models.
- Communicate and defend a scientific argument.

Science as Inquiry: The Understandings

Figure 4.3 summarizes the fundamental understandings that students should develop as a result of their science education.

Figure 4.3

Science as Inquiry: Fundamental Concepts for Grades 9–12

Fundamental Understanding About Scientific Inquiry

- Scientists usually inquire about how physical, living, or designed systems function.
- Scientists conduct investigations for a wide variety of reasons.
- Scientists rely on technology to enhance the gathering and manipulation of data.
- Mathematics is essential in scientific inquiry.
- Scientific explanations must adhere to criteria such as the following: A proposed explanation must be logically consistent; abide by the rules of evidence; be open to questions on possible modification; and be based on historical and current scientific knowledge.
- Results of scientific inquiry—new knowledge and methods—emerge from different types of investigations and public communication among scientists.

I turn to questions that emerge from the discussion of inquiry as content: "How do science teachers help students attain the abilities and understanding described in the Science as Inquiry Standards?" and "What do the Standards say about teaching?"

Science Teaching Standards

The science teaching standards (see Table 4.1) provide a comprehensive perspective for science teachers who wish to implement strategies that will provide students with the opportunities to experience science as inquiry. The national standards advocate the use of diverse teaching strategies to achieve varied outcomes. The *National Science Education Standards* state,

> *Although the standards emphasize inquiry, this should not be interpreted as recommending a single approach to science teaching. Teachers should use different strategies to develop the knowledge, understandings, and ability described in the content standards. Conducting hands-on science activities does not guarantee inquiry, nor is reading about science incompatible with inquiry. Attaining the understanding and abilities described in [the prior section] cannot be achieved by any single teaching strategy or learning experience.* (NRC 1996, pp. 23–24)

Table 4.1

Science Teaching Standards

Teaching Standard A Teachers of science plan an inquiry-based science program for their students.	In doing this, teachers • develop a framework of yearlong and short-term goals for students. • select science content and adapt and design curricula to meet the interests, knowledge, understanding, abilities, and experiences of students. • select teaching and assessment strategies that support the development of student understanding and nurture a community of science learners. • work together as colleagues within and across disciplines and grade levels.
Teaching Standard B Teachers of science guide and facilitate learning.	In doing this, teachers • focus and support inquiries while interacting with students. • orchestrate discourse among students about scientific ideas. • challenge students to accept and share responsibility for their own learning. • recognize and respond to student diversity and encourage all students to participate fully in science learning. • encourage and model the skills of scientific inquiry, as well as the curiosity, openness to new ideas and data, and skepticism that characterize science.
Teaching Standard C Teachers of science engage in ongoing assessment of their teaching and of student learning.	In doing this, teachers • use multiple methods and systematically gather data about student understanding and ability. • analyze assessment data to guide teaching. • guide students in self-assessment. • use student data, observations of teaching, and interactions with colleagues to reflect on and improve teaching practice. • use student data, observations of teaching, and interactions with colleagues to report student achievement and opportunities to learn to students, teachers, parents, policy makers, and the general public.
Teaching Standard D Teachers of science design and manage learning environments that provide students with the time, space, and resources needed for learning science.	In doing this, teachers • structure the time available so that students are able to engage in extended investigations. • create a setting for student work that is flexible and supportive of science inquiry. • ensure a safe working environment. • make the available science tools, materials, media, and technological resources accessible to students. • identify and use resources outside the school. • engage students in designing the learning environment.

Table 4.1 *(continued)*

Science Teaching Standards

Teaching Standard E Teachers of science develop communities of science learners that reflect the intellectual rigor of scientific inquiry and the attitudes and social values conducive to science learning.	In doing this, teachers • display and demand respect for the diverse ideas, skills, and experiences of all students. • enable students to have a significant voice in decisions about the content and context of their work and require students to take responsibility for the learning of all members of the community. • nurture collaboration among students. • structure and facilitate ongoing formal and informal discussion based on a shared understanding of rules of scientific discourse. • model and emphasize the skills, attitudes, and values of scientific inquiry.
Teaching Standard F Teachers of science actively participate in the ongoing planning and development of the school science program.	In doing this, teachers • plan and develop the school science program. • participate in decisions concerning the allocation of time and other resources to the science program. • participate fully in planning and implementing professional growth and development strategies for themselves and their colleagues.

The Essential Features of Inquiry in Science Classrooms

After publication of the *National Science Education Standards* (NRC 1996), we realized the need for an addendum that elaborated on inquiry, as it was a prominent feature of the standards. Work began on *Inquiry and the National Science Education Standards: A Guide for Teaching and Learning,* and the addendum was published in 2000. A key theme in this document was a description of the essential features of inquiry in the specific context of science classroom and science teaching. Following are the five essential features of inquiry.

1. Learner engages in scientifically oriented questions.
2. Learner gives priority to evidence in responding to questions.
3. Learner formulates explanations from evidence.
4. Learner connects explanations to scientific knowledge.
5. Learner communicates and justifies explanations.

The next sections describe the essential features in greater detail. These descriptions are adapted from the aforementioned addendum (NRC 2000).

Essential Feature 1: Learners Are Engaged by Scientifically Oriented Questions.

Scientifically oriented questions center on objects, organisms, and events in the natural world; they connect to the science concepts described in the content standards. These questions lend themselves to empirical investigation and lead to gathering and using data to develop explanations for scientific phenomena. Scientists recognize two primary kinds of scientific questions. Existence questions probe origins and include many "why" questions. Why do objects fall toward Earth? Why do some rocks contain crystals? Why do humans have chambered hearts? There also are causal and functional questions, which probe mechanisms and include *how* questions. How does sunlight contribute to plant growth? How are rocks formed?

In the classroom, a question can drive an inquiry and generate a need to know in students, stimulating additional questions about natural phenomenon. The initial question may originate from the learner, teacher, curriculum materials, internet, or other sources. The science teacher may play a critical role in guiding the identification of questions, particularly when they come from students. Fruitful inquiries develop from questions that are meaningful and relevant to students, but they also must be able to be answered by students' observations and the scientific knowledge they can obtain from reliable sources. The knowledge and procedures students use to answer the questions must be accessible and manageable, as well as appropriate to the students' developmental levels.

Essential Feature 2: Learners Give Priority to Evidence That Allows Them to Develop and Evaluate Explanations That Address Scientifically Oriented Questions.

Science distinguishes itself from other ways of knowing through the use of empirical evidence as the basis for explanations about the natural world. Scientists concentrate on getting accurate data from observations and experiments. They obtain evidence from getting accurate data from observations and experiments. They obtain evidence from observations and measurements taken in natural settings or in settings such as laboratories. They use their senses; instruments such as telescopes and microscopes to enhance their senses; and instruments that measure characteristics that humans cannot sense, such as magnetic fields. In some instances, scientists can control conditions to obtain their evidence; in other instances, they cannot control the conditions or control would distort the phenomena, so they gather data over a wide range of naturally occurring conditions and over a long enough period of time that they can infer the influence of different factors. The accuracy of the evidence gathered is verified by checking measurements, repeating the observations, or gathering different kinds of data related to the same phenomena. The evidence is subject to questioning and further investigation.

The above paragraph explains what counts as evidence in science. In their classroom inquiries, students use evidence to develop explanations for scientific phenomena. They observe plants, animals, and rocks and carefully describe their characteristics. They can take measurements of temperatures, distances, and time and carefully record them. They can observe chemical reactions, predator and prey relationships, and moon phases and chart their results and interactions. They also may obtain facts and information from their teacher, curriculum materials, or the internet to facilitate their inquiries.

Essential Feature 3: Learners Formulate Explanations From Evidence to Address Scientifically Oriented Questions.

This aspect of inquiry emphasizes the connection between evidence and explanation rather than the criteria for and characteristics of the evidence. Scientific explanations should be formulated using logic and reason. They provide causes for effects and establish relationships based on evidence and logical argument. They must be consistent with observational and experimental evidence about nature. Scientific explanations respect rules of evidence, are open to criticism, and require the use of various processes generally associated with science— for example, classification, analysis, inference, and prediction—and cognitive processes such as critical reasoning and logic.

Proposed explanations extend what is known to what is unknown, so explanations go beyond current knowledge and propose some new understanding. For science, this means building on extant knowledge. For students, this means expressing new ideas based on their current understandings. In both cases, the result is a proposed explanation.

Essential Feature 4: Learners Evaluate Their Explanations in Light of Alternative Explanations, Particularly Those Reflecting Scientific Understanding.

Evaluation, and possible elimination or revision of explanations, is one feature that distinguished scientific inquiry from other ways of knowing. Science teachers can ask questions such as, "Does the evidence support the proposed explanation?" "Does the explanation adequately answer the questions?" "Are there any apparent biases or flaws in the reasoning connecting evidence and explanation?" "Can other reasonable explanations be derived from the same evidence?"

Alternative explanations may be reviewed as students engage in dialogues, compare results, or check their results with those proposed by others. This characteristic ensures that students make the connection between their results and appropriate scientific knowledge. That is, given their age and stage of development, students' explanations should ultimately be consistent with currently accepted scientific knowledge.

Essential Feature 5: Learners Communicate and Justify Their Proposed Explanations.

Scientists communicate their explanations in such a way that their results can be reproduced. This requires clear articulation of the question, procedures, evidence, and proposed explanation, as well as a review of alternative explanations. It provides for further skeptical review and the opportunity for other scientists to use the explanation in working on new questions.

Having students share their explanations provides other students and teachers the opportunity to ask questions, examine evidence, identify faulty reasoning, point out statements that go beyond the evidence, and suggest alternative explanations for the same observations. Sharing explanations can bring into question or fortify the connections students have made among the evidence, existing scientific knowledge, and their proposed explanations. As a result, students can resolve contradictions and solidify an empirically based argument.

Variations of Inquiry in Science Classrooms

One of the unfortunate misconceptions about teaching science as inquiry is that all inquiry must originate with a student's question. In the extreme, this position does not allow for other origins for questions, such as the science teacher asking a question, conducting a demonstration, or engaging students in an activity. *Inquiry and the National Science Education Standards* (NRC 2000) presents a view of classroom inquiry that is not this either/or position. Rather, the view is one of a continuum and variations that may range from less to more self-direction by students and more to less direction by teachers, materials, or other sources. Table 4.2 (p. 86) uses the essential features of inquiry and presents this continuum.

Table 4.2

Essential Features of Classroom Inquiry and Their Variations Along a Continuum

More<---------------------------Amount of Learner Self-Direction--------------------------->Less Less<--------------------Amount of Direction From Teacher or Written Material----------------->More				
Learner ENGAGES in scientifically oriented questions.	Learner poses a question.	Learner selects among questions, poses new questions.	Learner sharpens or clarifies question provided by teachers, materials, or other source.	Learner engages in question provided by teacher, materials, or other source.
Learner gives priority to EVIDENCE in responding to questions.	Learner determines what constitutes evidence and collects it.	Learner directed to collect certain data.	Learner given data and asked to analyze.	Learner given data and told how to analyze.
Learner formulates EXPLANATIONS from evidence.	Learner formulates explanation after summarizing evidence.	Learner guided in process of formulating explanations from evidence.	Learner given possible ways to use evidence to formulate explanation.	Learner provided with evidence.
Learner connects explanations to scientific KNOWLEDGE.	Learner independently examines other resources and forms the links to explanations.	Learner directed toward areas and sources of scientific knowledge.	Learner given possible connections.	Learner given connections to scientific knowledge.
Learner COMMUNICATES AND JUSTIFIES explanations.	Learner forms reasonable and logical argument to communicate explanations.	Learner coached in development of communication.	Learner provided broad guidelines to use to sharpen communication.	Learner given steps and procedures for communication.

Source: National Research Council (NRC). 2000. *Inquiry and the national science education standards: A guide for teaching and learning.* Washington, DC: National Academies Press. p. 29.

Some Research Worth Noting

This section presents some research supporting the proposal to teach science as inquiry. I criticize a contemporary view supporting direct instruction and establish a linkage among research, instruction, and inquiry.

A Definition of Inquiry

Inquiry as presented in science education has several different and quite distinctive meanings. Beginning with a definition that I developed using a common dictionary form will help set the parameters for further discussion.

> In.quir.y In´ kwir´ ē) n., pl. ies. 1. *An outcome of science teaching that is characterized by* **knowledge** *and* **understanding** *of the processes and methods of science. 2. Outcomes of science teaching that refer to specific* **skills** *and* **abilities** *integral to the processes and methods of science. 3. The instructional strategies used to achieve students' knowledge and understanding of science* **concepts, principles,** *and* **facts** *and/or the outcomes described in the aforementioned definitions 1 and 2.*

This short statement differentiates between inquiry as teaching strategies and inquiry as the learning outcomes of a science teacher. The distinction between teaching strategy and learning outcome is not as clear as the headings indicate because teaching science as inquiry requires some use of inquiry-oriented strategies and inevitable results in learning outcomes associated with knowledge and understanding or skills and abilities.

Historically, there always have been individuals and groups advocating different strategies for teaching science. On one end of a continuum is direct instruction. Lecture serves as the example of this teaching method. At the other end of this continuum is full, unguided inquiry. The extreme position in this view is that students must discover scientific knowledge themselves without any guidance from the teacher. In reality, most science teaching is somewhere in the middle of the continuum. Effective science teaching embodies a variety of strategies and methods. One difficulty, however, is that terms such as *direct instruction* and *inquiry learning* often are argued from either/or positions.

The Inquiry Synthesis Project

The Education Development Center (EDC) in Boston completed an extensive review of qualitative and quantitative research on inquiry. Known as the "Inquiry Synthesis Project," the research team reviewed research between 1984 and 2002 to answer the central question of the project: What is the impact of inquiry science instruction on student outcomes?

Methodology for the project consisted of three phases: report collection, coding, and analysis. An initial review identified 443 research reports, of which

138 met the criteria for inclusion in the final analysis. The review and inclusion or exclusion of studies in the final analysis was among the most rigorous I have seen. The team established five components of inquiry science instruction. Those components included

- developing investigation questions,
- designing experiments,
- collecting data,
- drawing conclusions, and
- communicating results.

The dependent variable was retention of knowledge or understanding (i.e., facts, concepts, principles, theories) in the physical, life, and Earth sciences.

Several conclusions are worth noting. First, a majority (51%) of the studies showed positive results for inquiry-based science instruction on learning outcomes. Second, the research team completed a further analysis of comparative studies (i.e., quasi-experimental and experimental designs with comparison groups) and found that 63% of these studies demonstrated a statistically significant increase in students' understanding of science concepts for those who received higher levels of inquiry-based instructional experiences (EDC 2007; Minner, Levy, and Century 2010).

The EDC research team gave appropriate cautions about the conclusions of this synthesis and use of the results to declare a winner or loser in debates about inquiry instruction versus direct instruction. This said, I would give a slight advantage to inquiry-based instruction based on the rigorous methodology employed by the EDC team, the number of studies included, and the positive results. Considering the relationship to the definition described at the beginning of the section, I would note that the primary emphasis was on instructional strategies used to achieve knowledge and understanding of science concepts, principles, and facts in the physical, life, and Earth sciences.

Inquiry Strategies Versus Direct Instruction

Research headed by David Klahr has stimulated review and discussion of the relative importance of direct instruction and inquiry learning (Klahr has used the term *discovery learning*) as instructional approaches to science teaching (Chen and Klahr 1999; Klahr, Chen, and Toth 2001; Klahr and Li 2005; Klahr and Nigam 2004). In a 1999 study, Chen and Klahr investigated one important aspect of scientific reasoning. They asked the question, "What is the effectiveness of different instructional strategies in children's acquisition of the domain-general strategy—Control of Variables Strategy (CVS)?" They had children ages 7 to 10 years old design and evaluate experiments after direct instruction about CVS and without direct instruction—that is, inquiry learning in the extreme, unguided form. They reported that with direct instruction children did learn and could transfer the basic strategy for designing unconfounded experiments—that is, they could apply CVS (Chen and Klahr 1999). Before continuing this discussion of Klahr's research, I will introduce a report on the use of the laboratory in high school science. The report includes an important perspective on instruction that directly relates to this discussion. I will return later to Klahr's research.

In 2006, the National Research Council published *America's Lab Report: Investigations in High School Science* (NRC 2006). The NRC proposed the *phrase integrated instructional units* to describe the design of instructional units that carefully combine laboratory experiences with other types of teaching strategies, including lectures, reading, and discussion. Research indicates that integrated instructional units increase students' mastery of subject matter compared with other modes of instruction, and, very important, these units aid the development of more sophisticated aspects of scientific reasoning, increase students' interest in science, and somewhat improve students' understanding of the nature of science when this goal is explicitly targeted (NRC 2006, p. 100). All of these are valued goals of science education. Upon reading this research, I immediately made several connections. First, integrated instructional units had the design features of the BSCS 5E Instructional Model. Second, integrated instructional units were not exclusively "direct instruction" but may include direct instruction; they were not unguided inquiry but could include activities and strategies embodying the essential features of guided inquiry (NRC 2000). Third, both the NRC report and David Klahr's research claimed support for their respective strategies as being effective for the development of some aspects of scientific reasoning, which is a critical outcome of inquiry-based instruction.

The research methodology used by Klahr and his colleagues actually paralleled that of an instructional model or an integrated instructional unit. Although the varied teaching methods were evident in the articles, Klahr and colleagues concluded that direct instruction was the critical strategy. The following quotes are from the methodological section of one of the key articles cited in the direct instruction versus inquiry learning debate (Chen and Klahr 1999). In my view, the entire methodology could be described as an integrated instructional unit that centers on students learning the key concepts of the Control of Variables Strategy.

The present study consisted of two parts. Part I included hands-on design of experiments. Children were asked to set up experimental apparatus so as to test the possible effects of different variables. The hands-on study was further divided into four phases. In Phase 1, children were presented with materials in a source domain in which they performed an initial exploration followed by (for some groups) training. Then they were assessed in the same domain in Phase 2. In phases 3 and 4, children were presented with problems in two additional domains (Transfer-1 and Transfer-2). Part II was a paper-and-pencil posttest given two months after Part I. The posttest examined children's ability to transfer the strategy to remote situations. (Chen and Klahr 1999, p. 4)

David Klahr and his colleagues present a well-designed study that, I would argue, used an integrated instructional approach that closely resembles the BSCS 5E Instructional Model. With an engagement based on the orientation and hands-on introduction to materials, the researchers had the students continue with an exploration, proceed to an explanation and demonstration of CVS, then apply or elaborate CVS to new situations for which they used the terms *assessment*, *Transfer-1*, and *Transfer-2*.

One could reasonably argue that the research methods employed by Klahr and his colleagues used instructional sequences that integrated different strategies but then isolated one strategy, direct instruction, as the key factor in learning. Others have generalized these results to claim that direct instruction is the best way to teach the processes and methods of science and, in the extreme, all of science (Adelson 2004; Cavanagh 2004; Begley 2004a, 2004b). In my view, such extreme generalizations based on the methodology and data of the Klahr studies extend beyond the reasonable limits of the studies. However, the research does provide insights that may help answer questions about effective instructional strategies that could be identified as inquiry oriented.

"How does inquiry-based instruction contribute to the development of knowledge and skills for the 21st century?" This, it seems to me, is a reasonable and appropriate question. Answering the question may advance our understanding of the form and function of inquiry in science education. Based on recent reports from the National Research Council (Bransford, Brown, and Cocking 1999; Donovan and Bransford 2005; NRC 2006), I argue that using an integrated instructional sequence that incorporates varied teaching methods holds the key to a reasonable and appropriate approach to teaching science as inquiry.

The design of integrated instructional units requires the careful selection of activities on the basis of research-based ideas likely to enhance learning. Laboratory and other experiences are explicitly linked. As I mentioned earlier, the BSCS 5E Instructional Model meets the design criteria for integrated instructional units. The strategies used in such units may include direct instruction, discrepant events, laboratories, discussions, demonstrations, readings, debates, virtual field trips, and other activities and methods common to curriculum and instruction in science.

PISA 2006 and Instruction in Science

PISA 2006 emphasized science and included a test of scientific competencies and a student questionnaire that included different opportunities to learn, specifically questions about science lessons. Based on data from PISA 2006, colleagues in Germany completed an analysis and report on instruction in science (Seidel et al. 2008). The German research team analyzed PISA data for variables that were considered relevant for attaining scientific literacy. These variables include learning time, teacher-student interactions, experimenting with student-conducted research, and scientific modeling and applications. In addition, and important in this discussion, the German team analyzed lesson patterns rather than individual lesson characteristics. Here, I make the connection to "integrated instructional unit" (NRC 2006), guided inquiry (NRC 2000), and a specific example of these that has special meaning for me, the BSCS 5E Instructional Model.

The questions asked in this analysis were:

1. In which countries is the complex process of scientific thinking and scientific methods consistently and also frequently considered in everyday lessons?

The specific elements of this question should by now be familiar to the reader: In almost all lessons, students have the opportunities to (1) plan their own experiments; (2) carry out practical experiments; (3) draw conclusions from experiments; (4) discuss their own ideas and basis for explanations; and (5) recognize contributions that science makes to society (see summary in Figure 4.4).

Figure 4.4

Five Basic Elements Used as a Basis for Lesson Patterns

1. Students can develop their own experiments.

2. Students carry out practical experiments in the laboratory.

3. Students should draw conclusions from an experiment that they have carried out.

4. Students are given the opportunity to explain "their own" ideas.

5. Teacher uses science lessons to make the world outside of school comprehensible for students.

2. This question centers on "dosage," how much time and emphasis were given to different activities, methods, and learning approaches. To answer these questions, the research team combined five questions (listed in Figure 4.4) from the international questionnaire. These five questions represent a comprehensive lesson pattern that centers on scientific methods and scientific ways of thinking and the degree to which teachers provide structure and guidance in lessons. Analysis revealed three patterns of lessons that OECD countries have implemented.

The first pattern of lessons (referred to as Type 1) involves all five of the aforementioned methods of scientific experimenting and research in all or most lessons. I would characterize this lesson pattern as the extreme form of open or free inquiry. In OECD countries, 21% of students reported this pattern of lessons. The U.S. percentage for Type 1 was 29%.

The second pattern (Type 2) is characterized by less frequent opportunities to plan and carry out their own experiments. However, they regularly draw conclusions from experiments, explain their ideas, and apply science to the world outside of school. I suggest this is guided inquiry. In OECD countries, 45% of 15-year-olds reported this pattern. The U.S. percentage for Type 2 was 55%.

The third lesson pattern (Type 3) is characterized by the fact that the five characteristics of scientific experimenting and research are rarely encountered in lessons. In OECD countries, 34% of students report this pattern of lessons. The U.S. percentage for Type 3 was 16%.

Following are some conclusions of the German research team.

- Students who reported a lesson time of at least four hours per week achieve significantly higher levels of scientific literacy than students who have a weekly lesson time of less than two hours. This finding is true for all OECD countries. A difference in the U.S. level of scientific literacy is evident—a difference of 69 points on the assessment, which is above the OECD average of 62 points.

- Students who experience Type 2 patterns of lessons—less frequent opportunities to plan and carry out their own experiments but draw conclusions, explain ideas, and apply science to the world outside of school—clearly demonstrate higher levels of scientific literacy. Type 1 patterns of lessons result in the lowest performance scores. The score for Type 2 in the United States is 52 points above the average performance score for Type 1. Type 3 scores are above Type 1 but lower than Type 2.

- Students' interest in science varies with the patterns of lessons. Not surprisingly, Type 1 patterns result in the highest interest in science. On average for OECD countries, Type 2 is slightly below Type 1. This difference in the United States is greater than average.

This analysis suggests that students should have four hours or more of science per week, and from the view of multiple learning outcomes (i.e., scientific literacy and interest in science), the use of a Type 2 pattern of lessons would be ideal.

In the context of this chapter, secondary school science should include four hours or more per week and use integrated instructional units that are characterized as guided inquiry.

Table 4.3 presents linkages among the research of Klahr, the BSCS 5E Instructional Model, and the essential features of inquiry as described in the National Research Council report *Inquiry and the National Science Education Standards* (NRC 2000).

Table 4.3

Linking Research, Instruction, and Inquiry

Chen and Klahr (1999)	An Integrated Instruction Sequence (NRC 2006): The BSCS 5E Instructional Model	Essential Features of Inquiry (NRC 2000)
"Children were presented materials in a source domain in which they performed an initial exploration."	**Engagement** initiates the learning process and exposes students' current conceptions.	Teachers can engage learners with demonstrations, activities, and field trips to form the basis for scientifically oriented questions.
"Children were asked to set up experimental apparatus so as to test the possible effects of different variables."	In the **Explore** phase, students gain experience with phenomena or events.	Learners can use the results of laboratory investigations to give priority to evidence and to allow them to address scientific questions.
"… included an explanation of the rationale behind controlling variables as well as examples of how to make unconfounded comparisons."	In the **Explain** phase, the teacher may give an explanation to guide students toward a deeper understanding.	Learners formulate explanations and teachers can provide direct instruction about scientific concepts, principles, and facts.
"… children were presented with problems in two additional domains."	In the **Elaborate** phase, students apply their understanding in a new situation or context.	Learners evaluate scientific explanations as they apply them to new situations.
"Part II was a pencil-and-paper posttest given two months after Part I."	In the **Evaluate** phase, student understanding and transfer are assessed.	Learners communicate and justify their proposed scientific understanding.

Concluding Discussion

In conclusion, I have tried to bring some clarity to the term *inquiry* as it applies to school science programs and the preparation of young minds. First, teaching science as inquiry includes understanding scientific inquiry and developing the cognitive abilities associated with the processes and methods of science. Second, inquiry can refer to an integrated and linked instructional sequence designed with the intention of helping students learn science concepts, as well as understanding inquiry and developing cognitive abilities aligned with inquiry. It is past time to move beyond the old either/or arguments of inquiry versus direct instruction. Science teachers have always used multiple strategies, so we need not make a decision about the one best strategy for teaching science. There isn't one; there are many strategies that can be applied to achieve different outcomes. Science teachers should try to sequence the various strategies into coherent and focused ways. This is teaching science as inquiry.

Science Teaching and Assessing Students' Scientific Literacy

In this chapter, I introduce some dimensions of scientific literacy and describe PISA, the Program for International Student Assessment, as the basis for understanding scientific literacy from both teaching and assessment perspectives.

Most science educators agree that the purpose of school science is to help students achieve levels of scientific literacy. The following discussion answers these questions: What do we mean by scientific literacy? What does scientific literacy imply for curriculum and instruction? What counts as achieving scientific literacy?

In *Achieving Scientific Literacy: From Purpose to Practices* (Bybee 1997), I summarize a conceptual framework for scientific literacy. The framework is a threshold model, which assumes that scientific literacy is continuously distributed within the population. At one extreme one can identify a small number of scientifically illiterate individuals; then, and across the population, there is a distribution of individuals who demonstrate increasingly greater degrees of scientific literacy. At the other end of the distribution there exists a small number of individuals whose level of scientific literacy is extremely high. The model also accommodates different perspectives, such as science disciplines, history and nature of science, relationships between science and society, and so on. The degree of scientific literacy demonstrated by any individual at any one time is a function of a range of factors—age, developmental stage, life experiences, and quality of science education, which includes an individual's formal, informal, and incidental learning experiences. The model describes certain thresholds that separate degrees of scientific literacy. The framework likewise provides a larger model that is useful to those constructing school science programs or teaching science.

The conceptual framework also favors inclusion rather than exclusion. Some attempts to define scientific literacy assume an either/or perspective: One is scientifically literate or scientifically illiterate. A more productive definition recognizes that scientific literacy develops over a lifetime and that a majority of individuals can be described positively as demonstrating some degree of scientific literacy.

The framework also accommodates the fact that a person may, at any time, be compared to the population as a whole and may demonstrate several levels of literacy at once depending on the historical context, social issue, and science discipline. Likewise, subgroups of similar individuals, whether scientists or middle school students, may be located at different points on the scientific literacy continuum.

Scientific literacy is a continuum in which an individual develops greater and more sophisticated understanding of science. This framework also functions as a taxonomy for current programs and practices and as a guide for future curriculum development and instructional approaches.

Nominal Scientific Literacy

In *nominal* literacy, the individual associates names with a general area of science and technology. However, the association may represent a misconception, naïve theory, or everyday explanation. Using the basic definition of *nominal*, the relationship between science terms and acceptable definitions is small and insignificant. At best, students demonstrate only a token understanding of science concepts, one that bears little or no relationship to real understanding.

Functional Scientific Literacy

Individuals demonstrating a *functional* level of literacy respond adequately and appropriately to vocabulary associated with science and technology. They meet minimum standards of literacy as it is usually understood; that is, they can read and write passages with simple scientific and technological vocabulary. Individuals may also associate vocabulary with larger conceptual schemes—for example, that genetics is associated with variation within a species and variation is associated with evolution—but only have a token or marginal understanding of the associations.

Conceptual and Procedural Scientific Literacy

Conceptual and procedural literacy occurs when individuals demonstrate an understanding of both the parts and the whole of science as a discipline. The individual can identify the way the parts form a whole vis-à-vis major conceptual schemes and the way new explanations and inventions develop vis-à-vis the processes of science and technology. At this level, individuals understand the conceptual structure of disciplines and the methodological procedures for developing new knowledge.

Multidimensional Scientific Literacy

Multidimensional literacy consists of understanding the essential conceptual structures of science as well as the features that make that understanding more complete, such as the history and nature of science. In addition, individuals at this level understand the relationship of disciplines to the whole of science and technology and have competencies that contribute to the application of scientific knowledge to personal and societal situations.

Scientific literacy implies a general education as opposed to specific future scientist education for school programs. Although some science educators have written about science and general education, science educators as a community have not developed this idea as fully as it should be.

Scientific literacy is best defined as a continuum of understanding about the natural and the designed world, from nominal to functional, conceptual, and procedural, and multidimensional. This unique perspective broadens the concept of scientific literacy to accommodate all students and give direction to classroom teachers and those responsible for curriculum, assessment, research, professional development, and administration to a broad range of students.

PISA 2006 provided an opportunity to survey the scientific literacy of 15-year-olds in 57 countries that constitute approximately 90% of the world economy. The next sections introduce PISA and identify potential linkages among scientific literacy and classroom curriculum, instruction, and assessment.

PISA 2006: An Assessment of Scientific Literacy

The Programme for International Student Assessment (PISA) presents a unique perspective on the science education landscape. Most assessments look back at what students should have learned and whether they attained the knowledge and skills described in the science curriculum. This observation is true for most classroom, state, and national assessments. PISA looks ahead. Results from PISA are used to extrapolate students' present knowledge, attitudes, and skills to assess their potential abilities in the future. At age 15, how well can students apply their knowledge and skills in novel settings? The key point here is the ability students have to apply their knowledge and skills because that is what they will have to do in the future—as citizens. The essence and intended meaning of scientific literacy is similar to PISA's orientation.

The following sections introduce PISA 2006. This discussion is based on the publication *Assessing Scientific, Reading, and Mathematical Literacy: A Framework for PISA 2006* (OECD 2006). In addition, *PISA Science 2006: Implications for Science Teachers and Teaching* (Bybee and McCrae 2009) will interest science teachers.

PISA 2006: An Introduction

The Programme for International Student Assessment (PISA) measures 15-year-olds' capabilities in reading, mathematics, and science every three years. PISA was first implemented in 2000, and the most recent results are for the 2006 assessment. Each three-year cycle assesses one subject in depth. The other two subjects also are assessed, but not in the same depth as the primary domain. In 2006, science was the primary subject assessed. PISA also measures cross-curricular competencies. In 2003, for example, PISA assessed problem solving.

PISA is sponsored by the Organization for Economic Cooperation and Development (OECD), an intergovernmental organization of 30 industrialized nations based in Paris, France. In 2006, 57 countries participated in PISA—30 OECD countries and 27 non-OECD countries.

PISA uses the term *literacy* within each subject area to indicate a focus on the application of knowledge and abilities. As discussed in the prior section, *literacy* refers to a continuum of knowledge and abilities; it is not a typological classification of a condition that one has or does not have.

Scientific Literacy

For the purposes of PISA 2006, *scientific literacy* referred to an individual's scientific knowledge and use of that knowledge to *identify scientific questions,* to *explain scientific phenomena,* and to *draw evidence based conclusions* about science-related issues. In addition, the definition includes the understanding of the characteristic features of science as a form of human knowledge and inquiry; an awareness of how science and technology shape our material, intellectual, and cultural environments; and a willingness to engage in science-related issues.

The definition of scientific literacy provides for a continuum from less developed to more developed scientific literacy—that is, individuals are deemed to be more or less scientifically literate; they are not regarded as either scientifically literate or scientifically illiterate (Bybee 1997; Koballa, Kemp, and Evans 1997). So, for example, the student with less developed scientific literacy might be able to recall simple scientific factual knowledge and use common scientific knowledge to draw or evaluate conclusions. A student with more developed scientific literacy will demonstrate the ability to create or use conceptual models to make predictions or give explanations, formulate and communicate predictions and explanations with precision, analyze scientific investigations, relate data as evidence, evaluate alternative explanations of the same phenomena, and communicate explanations with precision.

The PISA 2006 definition of scientific literacy consists of four interrelated and complementary aspects:

- Recognizing life situations involving science and technology. This is the *context* for assessment units and items.

- Understanding the natural world, including technology, on the basis of scientific knowledge that includes both knowledge *of* the natural world and knowledge *about* science itself. This is the *knowledge component* of assessment.
- Demonstrating competencies that include identifying scientific questions, explaining phenomena scientifically, and using scientific evidence as the basis for arguments, conclusions, and decisions. This is the *competency component*.
- Responding with an interest in science, support for scientific inquiry, and motivation to act responsibly toward, for example, natural resources and environments. This is the *attitudinal dimension* of assessment.

This relationship is represented graphically in Figure 5.1.

Figure 5.1

Framework for PISA 2006 Science Assessment

Personal, social, and global contexts

- Life situations that involve science and technology

require you to

Scientific competencies

- identify scientific issues,
- explain phenomena scientifically, and
- use scientific evidence.

How you do so is influenced by

Scientific knowledge

What you know about

- the natural world (knowledge of *science*), and
- science itself (knowlege *about* science).

Attitudes toward science

- How you respond to science issues (interest, support for scientific inquiry, responsibility

Although this discussion and the framework in Figure 5.1 emphasize assessment, science teachers can make connections to science programs including curriculum materials and classroom instruction.

In the foreseeable future, citizens will have to address numerous challenges that are clearly related to science and technology. Citizens confront these and other challenges from perspectives that are personal, social, and global. The PISA 2006 Science survey used the contexts displayed in Table 5.1 for the assessment.

Table 5.1

Contexts for the PISA 2006 Science Assessments

	Personal (Self, Family, and Peer Groups)	Social (the Community)	Global (Life Across the World)
Health	Maintenance of health, accidents, nutrition	Control of disease, social transmission, food choices, community health	Epidemics, spread of infectious diseases
Natural Resources	Personal consumption of materials and energy	Maintenance of human populations, quality of life, security, production and distribution of food, energy supply	Renewable and nonrenewable, natural systems, population growth, sustainable use of species
Environment	Environmentally friendly behavior, use and disposal of materials	Population distribution, disposal of waste, environmental impact, local weather	Biodiversity, ecological sustainability, control of pollution, production and loss of soil
Hazards	Natural and human-induced hazards, decisions about housing	Rapid changes (earthquakes, severe weather), slow and progressive changes (coastal erosion, sedimentation), risk assessment	Climate change, effect of modern warfare
Frontiers of Science and Technology	Interest in science's explanations of natural phenomena, science-based hobbies, sport and leisure, music, and personal technology	New materials, devices and processes, genetic modification, weapons technology, transport	Extinction of species, exploration of space, origin and structure of the universe

Source: Organisation for Economic Co-operation and Development (OECD). 2006. *Assessing scientific, reading, and mathematical literacy: A framework for PISA 2006.* Paris: OECD.

The contexts in Table 5.1 were used in the PISA 2006 science survey. I include them here because they also could be the basis for curriculum and instruction directed toward scientific literacy.

Scientific Competencies

The PISA 2006 science assessment gave priority to the competencies listed in Table 5.2 (p. 102); the ability to identify scientifically oriented questions; describe, explain, or predict phenomena based on scientific knowledge; interpret evidence and conclusions; and use evidence to make and communicate decisions. These competencies involve scientific knowledge—both knowledge *of* science and knowledge *about* science.

Some cognitive processes have special meaning and relevance for scientific literacy. Among the *cognitive* processes that are implied in the scientific competencies are inductive/deductive reasoning, critical and integrated thinking, transforming representations (e.g., data to graphs, tables), constructing explanations and presenting an argument based on data, thinking in terms of models, and using mathematics.

Scientific Knowledge

Clear criteria were used to guide the selection of scientific knowledge that was assessed in PISA 2006. Moreover, the objective of PISA is to describe the extent to which students can *apply* their knowledge in contexts of relevance to their lives. The knowledge was selected from the major fields of physics, chemistry, biological science, and Earth and space science according to the following three criteria: relevance to real-life situations; importance for understanding physical, living, and Earth systems; and appropriateness to the development level of 15-year-olds.

In addition to assessing students' knowledge *of* science, PISA 2006 included assessments of students' knowledge and understanding of ideas *about* science, and of the interactions among science and technology and the material, intellectual, and cultural environments. The first category, "Scientific Inquiry," centers on inquiry as the central process of science and the various components of that process. Next is a category closely related to inquiry, that of "Scientific Explanations." Scientific explanations are the results of scientific inquiry. One can think of inquiry and explanations as the means of science (how scientists get data) and the goals of science (how scientists use data) as the basis for explanations of phenomena.

Attitudes

One goal of science teaching is students' development of attitudes that support their attending to scientific issues and the subsequent acquisition and application of scientific and technological knowledge for personal, social, and global benefits.

The PISA 2006 science assessment evaluated students' attitudes in three areas: interest in science, support for scientific inquiry, and responsibility for sustainable development. These areas were selected because they will provide an international portrait of students' general appreciation of science, their specific scientific attitudes and values, and their responsibility toward select science-related issues that have national and international ramifications. Note that this is not an assessment of students' attitudes toward school science programs or teachers. The results will provide information about the emerging problem of declining interest for science studies among young people.

Table 5.2 provides a summary of key components of the PISA 2006 science assessment.

Table 5.2

Summary of the Assessment Areas for PISA 2006—Science

Assessment Area	Description
Scientific literacy and its distinctive features	Scientific literacy refers to an individual's • Scientific knowledge and use of that knowledge to identify scientific issues, to explain scientific phenomena, and to use scientific evidence; • Understanding the characteristic features of science as a form of human knowledge and inquiry; • Awareness of how science and technology shape our material, intellectual, and cultural environments; and • Willingness to engage in science-related issues, and with the ideas of science, as a reflective citizen.
Science content	Areas of scientific knowledge and concepts include: • Physical systems • Living systems • Earth and space systems • Technological systems And knowledge about science, which includes: • Scientific inquiry • Scientific explanations
Scientific competencies	• Identify scientific questions • Explain phenomena scientifically • Use scientific evidence
Personal, social, and global contexts	Areas of application within the contexts include: • Health • Resources • Environments • Hazards • Frontiers of science and technology
Attitudes	The response to scientific situations include: • Interest in science • Support for scientific inquiry • Responsibility for sustainable development

Source: Organisation for Economic Cooperation and Development (OECD). 2007. *PISA 2006: Science competencies for tomorrow's world.* Danvers, MA: OECD.

PISA 2006 Science: An Overview of Results

The first part of this section presents the average scores for both OECD and non-OECD countries (Table 5.3). These results are presented in order to provide a larger view and locate the U.S. among the countries that participated, many of which are our economic competitors.

Table 5.3

PISA 2006 Science Literacy Scores by Country

Country	Mean Science Score (Standard Error)	Country	Mean Science Score (Standard Error)
Finland	563 (2.0)	United States	489 (4.2)
Hong Kong-China	542 (2.5)	*Lithuania*	488 (2.8)
Canada	534 (2.0)	Slovak Republic	488 (2.6)
Chinese Taipei	532 (3.6)	Spain	488 (2.6)
Estonia	531 (2.5)	Norway	487 (3.1)
Japan	531 (3.4)	Luxembourg	486 (1.1)
New Zealand	530 (2.7)	*Russian Federation*	479 (3.7)
Australia	527 (2.3)	Italy	475 (2.0)
Netherlands	525 (2.7)	Portugal	474 (3.0)
South Korea	522 (3.4)	Greece	473 (3.2)
Liechtenstein	522 (4.1)	*Israel*	454 (3.7)
Slovenia	519 (1.1)	*Chile*	438 (4.3)
Germany	516 (3.8)	*Serbia*	436 (3.0)
United Kingdom	515 (2.3)	*Bulgaria*	434 (6.1)
Czech Republic	513 (3.5)	*Uruguay*	428 (2.7)
Switzerland	512 (3.2)	Turkey	424 (3.8)
Austria	511 (3.9)	*Jordan*	422 (2.8)
Macao-China	511 (1.1)	*Thailand*	421 (2.1)
Belgium	510 (2.5)	*Romania*	418 (4.2)
Ireland	508 (3.2)	*Montenegro*	412 (1.1)
Hungary	504 (2.7)	Mexico	410 (2.7)
Sweden	503 (2.4)	*Indonesia*	393 (5.7)
OECD average	**500 (0.5)**	*Argentina*	391 (6.1)
Poland	498 (2.3)	*Brazil*	390 (2.8)
Denmark	496 (3.1)	*Columbia*	388 (3.4)
France	495 (3.4)	*Tunisia*	386 (3.0)
Croatia	493 (2.4)	*Azerbaijan*	382 (2.8)
Iceland	491 (1.6)	*Qatar*	349 (0.9)
Latvia	490 (3.0)	*Kyrgyzstan*	322 (2.9)

Non-OECD countries are shown in italics.
Source: Organisation for Economic Cooperation and Development (OECD). 2007. PISA 2006: Science competencies for tomorrow's world. Danvers, MA: OECD.

U.S. Students

How do U.S. students score on scientific literacy compared to 15-year-olds in other OECD and non-OECD countries? In 2006, the average U.S. student score was 489, compared to the OECD average of 500. Sixteen OECD countries had scores that were measurably higher than U.S. students. Top-performing countries included Finland (563), Canada (534), Japan (531), New Zealand (530), and Australia (527). In a ranking of OECD countries by scores, the United States was 21st (see Table 5.3).

Referring to non-OECD countries or jurisdictions, there were six countries with measurably higher scores than U.S. students. Top performing non-OECD countries included: Hong Kong (542), Chinese Taipei (532), Estonia (531), Liechtenstein (522), Slovenia (519), and Macao (511) (see Table 5.3).

Strengths of U.S. students included the scientific competencies: identification of scientific issues, knowledge about science (i.e., scientific inquiry and scientific explanations), and knowledge of Earth and space systems. U.S. students were weak in the following competencies: explaining phenomena scientifically and using scientific evidence. Students also were weak in their knowledge of living and physical systems.

In the next sections, I turn to more detailed results concerning one important context for teaching and assessing scientific literacy—the environment. These results include students' awareness, performance, concern, optimism, and responsibility, pertaining to environmental issues. There is a note of caution because students in different countries may have interpreted the questions in various ways.

PISA 2006: A Unique Approach to Science Literacy Assessment

Most school programs emphasize fundamental knowledge and processes of the science disciplines. These science programs are implicitly intended to provide students with the foundation for professional careers as scientists and engineers. This goal differs from that of scientific literacy as described in this chapter. With the centrality of science and technology to contemporary life, full participation in society requires that all adults, including those aspiring to careers as scientists and engineers, be scientifically literate. Classroom, state, and national curriculum instruction and assessments should reflect the goal of scientific literacy.

A Perspective on Assessing Scientific Literacy

When thinking about what science and technology might mean for individual citizens, one must consider several things simultaneously: the contexts within which citizens encounter science and technology, the extent to which they possess scientific knowledge, their understanding of science as a way of knowing, their attitudes toward science, and the extent to which they can bring the content and attitudes together to competently respond to life situations.

As opposed to assessments of science content, such as most state, national (NAEP), and international (TIMSS) assessments, PISA presents students with problems set in contexts that represent actual situations that one may encounter in life. PISA science units are developed in such a way that understanding science would not only benefit but also be required to adequately address the issues.

PISA begins with a perspective that centers on the capacity of students to *identify scientific issues, explain phenomena scientifically,* and *use scientific evidence* as they encounter, interpret, solve, and make decisions in life situations that involve science and technology. This discussion frames a central point of the PISA 2006 science assessment: the assessment focused on what 15-year-olds know, value, and are able to do within reasonable and appropriate personal, social, and global contexts that involve environmental issues.

The term *scientific literacy* may be emphasized for the following reasons: It is recognized as representing the goals of science education that should apply to all students; it connotes a broadness and an applied nature to the purposes of science education; it represents a continuum of scientific knowledge and the cognitive abilities associated with scientific inquiry; it incorporates multiple dimensions; and it incorporates the relationships between science and technology. It also is important to note that PISA 2006 includes attitudes, as they are essential to understanding the response to science-related life situations. For more information about the design and results of PISA 2006 Science, see the following resources: Bybee 2008; Bybee and McCrae 2009; Bybee, McCrae, and Laurie 2009; OECD 2009). The next sections present examples of assessment units from PISA 2006 Science. These discussions provide greater detail about assessing components of scientific literacy.

Assessment Units From PISA 2006

The appendix displays two examples of units from PISA 2006 Science. The units demonstrate environmental contexts, and the narratives report on competencies and levels of proficiency for scientific literacy.

Proficiency Levels in Science

Student scores in science for PISA 2006 were grouped into six proficiency levels. The six proficiency levels represented groups of tasks of ascending difficulty, with Level 6 as the highest and Level 1 as the lowest. The grouping into proficiency levels was undertaken on the basis of substantive considerations relating to the nature of the underlying competencies.

Table 5.4 (p. 106) is a map of science questions from the two examples, illustrating the proficiency levels and scientific competencies.

Table 5.4

A Map of Two Environmental Examples From PISA 2006

| Level | Lower Score Limit | Competency | | |
		Identifying Scientific Issues	Explaining Phenomena Scientifically	Using Scientific Evidence
6	707.9	ACID RAIN *Question 5.2 (717)* (full credit)	GREENHOUSE *Question 5 (709)*	
5	633.3			GREENHOUSE *Question 4.2 (659)* (full credit)
4	558.7			GREENHOUSE Question 4.1 (568) (partial credit)
3	484.1	ACID RAIN *Question 5.1 (513)* (partial credit)	ACID RAIN *Question 2 (506)*	GREENHOUSE *Question 3 (529)*
2	409.5			ACID RAIN *Question 3 (460)* (has embedded attitude item)
1	334.9			

Source: Organisation for Economic Cooperation and Development (OECD). 2007. *PISA 2006: Science competencies for tomorrow's world.* Danvers, MA: OECD.

Characteristics of the items within assessment units provide the basis for interpreting students' performance at different levels of proficiency and for different scientific competencies. The unit Acid Rain, for example, has questions that can be scored at Proficiency Levels 2, 3, and 6 and for all three competencies. Greenhouse has questions at Levels 3, 4, 5, and 6 and for the scientific competencies Explaining Phenomena Scientifically and Using Scientific Evidence.

At the very bottom of the scale, Proficiency Level 1 (below Level 2, the cut-point) for the competency, students must simply recall information. For example, students might be required to know that fossils of organisms were deposited at an earlier age and that active muscles get an increased flow of blood. At Proficiency Level 2, above the cut-point for the competency, students might be required to know the fact that freezing water expands and thus may influence the weathering of rocks. An example for the competency Using Scientific Evidence is Question 3 in Acid Rain. This question provides a good example for Proficiency Level 2. The item asks students to use information provided to draw a conclusion about the effects of vinegar on marble, a simple model for the influence of acid rain on marble.

For the lower levels of proficiency, items are set in simple and relatively familiar contexts and require only the most limited interpretation of a situation. Items only require direct application of scientific knowledge and an understanding of well-known scientific processes of science in familiar situations.

Around the middle of the proficiency scale, items require substantially more interpretation, frequently in situations that are relatively unfamiliar. Items often demand the use of knowledge from different scientific disciplines, including more formal scientific or technological representation, and the thoughtful linking of those different knowledge domains to promote understanding and facilitate analysis. They often involve a chain of reasoning or a synthesis of knowledge and can require students to express reasoning through a simple explanation. Typical activities include interpreting aspects of a scientific investigation, explaining certain procedures used in an experiment, providing evidence-based reasons for a recommendation, and identifying the origins of chemical elements in the atmosphere. In the unit Acid Rain, for example, students were provided information about the effects of vinegar on marble (i.e., a model for the effect of acid rain on marble) and asked to explain why some chips were placed in pure (distilled) water overnight. For partial credit at Proficiency Level 3, they simply had to state that it was a comparison. Level 6, for example, required them to state that the acid (vinegar) was necessary for the reaction. These responses were for the competency Identifying Scientific Issues.

For the competency Explaining Phenomena Scientifically, Acid Rain Question 2 provides an example. Here, students are asked about the origin of certain chemicals in the air. Correct responses required students to demonstrate an understanding of the chemicals as originating as car exhaust, factory emission, and burning fossil fuels.

For the competency Using Scientific Evidence, the unit Greenhouse presents a good example for Proficiency Level 3 in Greenhouse Question 3, students must interpret evidence, presented in graph form, and conclude that the combined graphs support a conclusion that both average temperature and carbon dioxide emission are increasing.

At the top of the proficiency scale, items typically involve a number of different elements, requiring even higher levels of interpretation. The selections are unfamiliar to students and require some degree of reflection and review. Items demand careful analysis, may involve more than a scientific explanation, and require carefully constructed arguments.

Typical items near the top of the scale involve interpreting complex and unfamiliar data, imposing a scientific explanation on a complex situation, and applying scientific processes to unfamiliar problems. At this part of the scale, items tend to have several scientific or technological elements that need to be linked by students, and their successful synthesis requires several interrelated steps. The construction of evidence-based arguments and communications also requires critical thinking and abstract reasoning.

An example for Proficiency Level 6 and the competency Explaining Scientific Phenomena is Question 5 of Greenhouse. Students must analyze a conclusion to account for other factors that could influence the greenhouse effect. A final example from Greenhouse centers on the competency Using Scientific Evidence and asks students to identify a portion of a graph that does not provide evidence supporting a conclusion. Students must locate a portion of two graphs where curves are not both ascending or descending and provide this finding as part of a justification for a conclusion.

Acid Rain serves as an example of a science unit containing embedded questions that query students' attitudes. Question 10N in Acid Rain probes the level of students' interest in the topic of acid rain, and Question 10S asks students how much they agree with statements supporting further research.

Students' Knowledge About and Attitudes Toward Environmental and Resource Issues

Responses to Cognitive Items About Resources and Environments

Through review and classification of PISA 2006 units in terms of natural resources and environments, we identified 10 units and a total of 32 items. Approximately one-third of the contextual situations on PISA 2006 Science included resources and environments. Proficiency levels for the items ranged from 1 through 6, with the majority of items at Levels 3 and 4. Acid Rain and Greenhouse serve as examples (see the appendix). For U.S. students, the average correct percentage overall for the environment and resource items versus the average percentage on the remaining 71 items was 46.74% versus 52.76%. The comparative percentages for OECD countries were 50.25% versus 54.56%. The United States did not perform as well, but this may be a reflection of the type of question.

We also looked at the percentage of students at each proficiency level who responded correctly to the items. For example, for Question 5 on Greenhouse: What percentage of Level 6 students answered this correctly? (Answer: 53%) Another example from Acid Rain: What percentage of students at Level 4 answered this question correctly? (Answer: 51%)

U.S. Students' Awareness of Environmental Issues

PISA 2006 surveyed students' awareness of selected environmental issues. As you can see in Table 5.5, the majority of U.S. students (73%) reported being aware of the consequences of clearing forest for other land uses. This percentage was the same as the OECD average. Just more than half of U.S. students are aware of acid rain, the increase of greenhouse gases in the atmosphere, and nuclear waste.

More than one-third of U.S. students (39%) are aware of the use of genetically modified organisms. This is higher than the OECD average, which was 35%.

Data from this survey also suggest that students' levels of awareness of environmental issues are strongly associated with their scientific knowledge. However, the United States was one country with a lower mean score in science (489) and students who are more aware of environmental issues. This linkage was true for all participating countries. Conversely, relatively lower scores on scientific knowledge may result in environmental issues being unnoticed, ignored, or dismissed by some citizens.

All students from more advantaged socioeconomic backgrounds reported higher levels of socioeconomic awareness of environmental issues. The United States and 24 of 30 other OECD countries also had a significant gender difference in students' awareness of environmental issues with boys scoring higher than girls.

U.S. Students' Levels of Concern About Environmental Issues

It is one thing to be aware and another to be concerned about environmental issues. PISA 2006 explored the latter by asking students to report whether or not selected issues were a serious concern to them and other people in their country. Students are, in general, concerned about global issues. As you can see in Table 5.6 (p. 110), the percentages are highest for air pollution (91% in the United States and 92% on average for OECD) and lowest for water shortage (81% in the United States and 76% for OECD).

Table 5.5

Students' Awareness of Selected Environmental Issues

Environmental Issue	Percentage of OECD Students Who Are Familiar With or Know Something About This Environmental Issue	Percentage of U.S. Students Who Are Familiar With or Know Something About This Environmental Issue
The consequences of clearing forests for other land use	73	73
Acid rain	60	54
The increase of greenhouse gases in the atmosphere	58	53
Nuclear waste	53	51
Use of genetically modified organisms (GMOs)	35	39

Table 5.6

Students' Level of Concern Regarding Environmental Issues

Environmental Issues	Percentage of OECD Students Who Believe the Following Environmental Issues to Be a Serious Concern for Themselves or Other People in Their Country	Percentage of U.S. Students Who Believe the Following Environmental Issues to Be a Serious Concern for Themselves or Other People in Their Country
Energy shortage	82	84
Water shortage	76	81
Air pollution	92	91
Nuclear waste	78	83
Extinction of plants and animals	84	85
Clearing of forests for other land use	83	87

Table 5.7

Students' Level of Optimism Regarding Environmental Issues

Environmental Issue	Percentage of OECD Students Who Believe the Following Environmental Issues Will Improve During the Next 20 Years	Percentage of U.S. Students Who Believe the Following Environmental Issues Will Improve During the Next 20 Years
Energy shortage	21	26
Water shortage	18	22
Air pollution	16	21
Nuclear waste	15	17
Extinction of plants and animals	14	18
Clearing of forests for other land use	13	15

Table 5.8

Students' Responsibility for Sustainable Development

Statements Describing Possible Policies on Student Questionnaire
A. Industries should be required to prove that they safely dispose of dangerous waste material.
B. I am in favor of having laws that protect the habitats of endangered species.
C. It is important to carry out regular checks on the emissions from cars as a condition of their use.
D. To reduce waste, the use of plastic packaging should be kept to a minimum.
E. Electricity should be produced from renewable resources as much as possible, even if this increases the cost.
F. It disturbs me when energy is wasted through the unnecessary use of electrical appliances.
G. I am in favor of having laws that regulate factory emissions even if this would increase the price of products.

Abbreviated Policy Statements Indicating Students' Responsibility	Percentage of OECD Students Who Strongly Agree With the Statement	Percentage of U.S. Students Who Strongly Agree With the Statement
A (Require safe disposal of waste)	92	88
B (Laws to protect habitats of endangered species)	92	90
C (Regular checks on car emissions)	91	89
D Minimize use of plastic packaging)	82	77
E (Produce electricity from renewable resources)	79	75
F (Wasted energy through unnecessary use of appliances)	69	63
G (Laws to regulate factory emissions)	69	56

In somewhat of a contrast with students' awareness, level of concern does not have a strong association with students' performance in science. Furthermore, students' levels of concern are not strongly associated with socioeconomic background. That is, students from less advantaged backgrounds are equally if not more concerned about environmental issues. That said, it also is the case they are less able to explain the issues. Finally, there is a significant gender difference in 29 of 30 OECD countries, with girls indicating greater concern than boys about environmental issues.

U.S. Students' Optimism Regarding Environmental Issues

To guide students' optimism about the future, PISA 2006 presented the same environmental issues from Table 5.6 and asked students if they thought the problems would improve during the next 20 years (see Table 5.7, p. 110). Only a minority of students in the United States and OECD countries thought the various environmental issues would improve within the next 20 years. U.S. students are most optimistic about solving the shortages of energy and water (26% and 22%, respectively). But about three-quarters of U.S. students are pessimistic about these two issues. Their optimism about other issues is even lower. Unfortunately, the association between science performance and optimism is weak to moderate. That is, the more students know about science, the less optimistic they seem to be. These results are similar to those found in the ROSE study (Schreiner and Sjoberg 2004).

Students from more disadvantaged socioeconomic backgrounds tend to be more optimistic about the improvement of these environmental issues within the next 20 years. Quite strikingly, girls are significantly less optimistic in 28 of 30 OECD countries, including the United States.

U.S. Students' Responsibility for Sustainable Development

If 15-year-old students express a general awareness, concern, and pessimism about environmental issues, it seems reasonable to ask about their sense of responsibility for sustainable development. PISA 2006 presented students with seven possible policies for sustainable development and asked them to respond by indicating the degree to which they agreed or disagreed with the policies. Students who indicated they agreed or strongly agreed were deemed to express a sense of responsibility for sustainable development. The strongest sense of responsibility was expressed for laws to protect endangered species (90% for the United States and 92% for OECD countries), followed by regular checks on car emissions (89% for the United States and 91% for OECD countries), and safe disposal of dangerous waste material (88% for the United States and 92% for OECD countries) (see Table 5.8, p. 111).

Here again, higher science performance is associated with a stronger sense of responsibility in all OECD countries. In general, students from more advantaged socioeconomic backgrounds tended to indicate a higher sense of responsibility for sustainable development. Interestingly, girls show significantly more responsibility than boys in 20 of 20 OECD countries, including the United States.

In conclusion, the results from PISA 2006 suggest that in general students with a greater understanding of science are also more aware of environmental issues. They also have a deeper sense of responsibility for sustainable development. However, these same students are not optimistic about how select environmental issues will improve during the next 20 years. Within this conclusion, boys tend to be more optimistic and girls tend to be more concerned and responsible about environmental issues.

Concluding Discussion

In an early section of the chapter, I described different dimensions of scientific literacy. Nominal literacy is the dimension in which students associate terms with a general area of science, but their general understanding is lacking overall.

Another dimension of scientific literacy includes vocabulary—the technical words of science and technology. I refer to this dimension as *functional scientific literacy*. Learners demonstrating functional scientific literacy use scientific words appropriately and adequately. Relative to science and technology, learners should meet minimum standards of literacy as it is usually defined. That is, given their age, stage of development, and education levels, learners should be able to read and write passages that include scientific and technological vocabulary.

For years, functional scientific literacy has received extraordinary emphasis in science teaching, and educators generally equated the goal of achieving scientific literacy with attaining vocabulary. Science teachers should reduce (but not eliminate) the current overemphasis on functional scientific literacy and increase the emphasis on other domains and dimensions of scientific literacy.

Conceptual and procedural scientific literacy describes another dimension of scientific literacy. Learners should relate information and experiences to conceptual ideas that unify the disciplines and fields of science. In addition, literacy in science must also include abilities and understandings relative to the procedures and processes that make science a unique way of knowing.

Scientific literacy extends beyond vocabulary, conceptual schemes, and procedural methods to include other understandings about science. We must help learners develop perspectives of science that include the history of scientific ideas, the nature of science and technology, and the role of science and technology in personal life and society. This is *multidimensional scientific literacy*.

The latter portion of the chapter used the PISA 2006 Science survey to explore the presentation of an assessment of scientific literacy. The framework, sample items, and results were examples that science teachers might use to design programs of curriculum, instruction, and assessment that might be used to answer the Sisyphean question in science education: What is important for citizens to know, value, and be able to do in situations involving science and technology?

Fulfilling National Aspirations Through Curriculum Reform

The science education community has a long history of responding to the needs and aspirations of society. This chapter and the next provide contemporary perspectives on the response of science education to national priorities and goals. I first explore this theme with reflections from the Sputnik era of curriculum reform. Reflecting on the Sputnik era provides insights about the ways and means in which the science education community responds to national goals. I use insights from this exploration to make recommendations that will help science teachers respond to aspirations for 21st-century science education.

The theme of sustaining global environments and conserving natural resources centers on curricular topics such as climate change and energy efficiency. For science teachers, these topics imply greater emphasis on students' understanding science and technology in personal, social, and global contexts.

Reflections From the Sputnik Era

The Sputnik era began in the 1950s with development of new programs that eventually became known by their acronyms. Science programs included the Physical Science Study Committee, known as PSSC Physics; the Chemical Educational Materials Study, known as CHEM Study; the Biological Sciences Curriculum Study, known as BSCS biology; and the Earth Sciences Curriculum Project, known as ESCP Earth science. At the elementary level there were the Elementary Science Study, known as ESS; the Science Curriculum Improvement Study, known as SCIS; and Science-A Process Approach, known as S-APA.

The years after World War II witnessed a debate between those supporting the progressive education identified with John Dewey and a conservative and traditional education identified with critics such as Admiral Hyman Rickover and Arthur Bestor. In fall 1957, the debate about American education reached a turning point. Sputnik resolved the debate in favor of traditionalists who recommended greater emphasis on higher academic standards, especially in science and mathematics. Sputnik made clear to the American public that it was in its national interest to change education, in particular the curriculum in mathematics and science. Although they had previously opposed federal aid to schools—on the grounds that federal aid would lead to federal control—the public required a change in American education. After Sputnik, the public demand for a federal response was unusually high, and Congress passed the National Defense Education Act (NDEA) in 1958.

Curriculum reformers of the Sputnik era shared a common vision and general plan of action. Across disciplines and within the education community, reformers generated significant national enthusiasm for their initiatives. They would replace the current content of facts and topics that had a progressive orientation with curricula based on the conceptually fundamental ideas of science and the modes of scientific inquiry, technological design, and mathematical problem solving. The reform would replace textbooks with instructional materials that included films, activities, and laboratories. No longer would schools' science, technology, and mathematics programs emphasize memorization of terms and applications of content. Rather, students would learn the conceptual structures and methodological procedures of science and mathematics disciplines.

These themes of curriculum reform should sound familiar. They are similar to those we have heard in the first decade of the 21st century. The context may be different (i.e., economic security, environmental quality, resource use), but the changes in curricula (i.e., fewer facts, core concepts, modes of inquiry, social connections) all reflect earlier themes.

The reformers' vision of replacing the curriculum, combined with united political support for education improvement, stimulated a reform that clearly centered on national goals. The Eisenhower administration (1953–1961) provided initial economic support, and the enthusiasm of the Kennedy administration (1961–1963) moved the nation forward on reform initiatives. Although the Soviet Union had provided Sputnik as a symbol for the problem, President Kennedy provided a manned flight to the moon by the end of the decade as America's vision, timeline, and strategic plan to win the race to space.

Reformers enjoyed financial support from both public and private sources for their curriculum projects. Federal agencies, particularly the National Science Foundation (NSF), and major philanthropic foundations, particularly Carnegie Corporation of New York and the Rockefeller Brothers Fund, provided ample support for attaining the vision, primarily through development of new curriculum programs.

The Sputnik era continued for two decades, into the mid-1970s. If I had to indicate an official end for the era, it would be 1976. *Man-A Course of Study* (MACOS), an anthropology program developed with NSF funds, came under scrutiny and widespread attack from conservative critics who objected to the subject matter (Dow 1991). When a House subcommittee held hearings, NSF conducted an internal review, and the Government Accounting Office (GAO) investigated the financial relationships between NSF and the developers, the end of the era was imminent.

The common vision was to improve school programs and provide students with background and experiences that would encourage their entry into scientific and technological arenas. In addition, the reforms tried to design "teacher-proof" curriculum materials—that is, materials that were so well designed that students would learn regardless of the teacher. Although there are numerous political and economic factors that provided countervailing influences on the reform, emphasis on educating students for future careers in science and developing "teacher-proof" programs contributed to a less-than-effective reform.

Five Insights From the Sputnik Era

Examination of the Sputnik era reveals some strength and weaknesses that are worth noting by contemporary reformers. I present several observations from the experience.

Curriculum Reform Is at Best Difficult, and Emphasizing "Teacher-Proof" Materials Makes It Impossible.

Replacement of school science programs is extremely difficult at best. Although leaders in the Sputnik era used terms such as *revision* and *reform*, the intention was to replace school science programs with teacher-proof curriculum materials. They had tremendous zeal and confidence for this ill-informed and misguided goal. They approached their programs and the reform with a "field of dreams" perspective. If they built good curriculum materials, then science teachers would adopt them, thus replacing traditional programs with teacher-proof curricula. Such an approach, however, faces pervasive institutional resistance, raises the personal concerns if not resentment of science teachers, alarms the public, and fails to recognize the essential role and responsibility of science teachers. Furthermore, the perspective does not acknowledge the systemic nature of science education.

From a science teacher's point of view, curriculum materials present a natural means of educational reform. Instructional materials are the means that science teachers use to improve their programs. There is a second insight embedded in this point. In some cases, entire programs are replaced when, for example, states or school districts adopt new elementary programs or high school science programs. With this type of reform, the selection of programs becomes a critical leverage point for professional development.

The lesson here centers on the importance of both recognizing the essential place of science teachers and supporting their work with a systemic approach to reform. Recall the instructional core theme. Not only are new programs important, but other components of the education system must change as well and provide support for the implementation of innovative programs and instructional practices. Those components include peers who are practicing science teachers; administrators; school boards; the community; and a variety of local, state, and national policies.

Resistance to Curriculum Reform Increases Proportionally to Variance From Current Programs.

The reluctance of science teachers to embrace and implement innovative programs and practices increases proportionally to the variance from current programs and practices. Teachers had difficulty with the content and pedagogy of new programs such as PSSC, BSCS, CHEM Study, SCIS, and ESS. Lacking educational support within their local systems and experiencing political criticism from outside education, teachers sought security by staying with or returning to the traditional programs.

The education lesson here centers on the importance of both initial and ongoing professional development and support for new knowledge and skills. In addition, education reformers have to recognize that changes in social and political forces have an effect on school programs. The importance of high-quality, sustained professional development aligned with curriculum reform cannot be overstated and is worth repeating. It is way past time to move beyond handing science teachers a new book and a single workshop and calling this curriculum reform and professional development.

Excluding Professionals in the Science Education Community Reduces Effectiveness.

The exclusion of other professionals in the larger science education community (e.g., teacher educators and science education researchers) contributed to a slower-than-desired acceptance of the new programs, reduced understanding by those entering the profession, and provided less-than-adequate and appropriate professional development for teachers in the classroom.

This is a lesson of professional inclusion. Education is a system consisting of many different components. One important component includes those who have some responsibility for teacher preparation, workshops and professional development, and the implementation of school science programs. Another important component involves assessment practices. A perspective that attempts to unify and coordinate efforts among teachers, educators, and scientists works best.

In contemporary education, the particular role of curriculum development groups often is undervalued or overlooked by states and school districts interested in improving their science programs. With the Sputnik era, the science education community created professional groups with the expertise to design, develop, and implement innovative, state-of-the-art curriculum materials.

Individuals working in these groups understand their role in bridging the gap between theory and practice. The combined expertise of those involved in curriculum reform stands in stark contrast to most materials designed by traditional textbook authors and teachers tasked with the development of local science programs.

Don't Underestimate State and Local Realities.

Realities and power of state and local school district policies, programs, and practices generally went unrecognized in the Sputnik era. Support from federal agencies and national foundations freed developers from the political and educational constraints of state and local agencies and the power and influence of commercial publishers.

This lesson directs attention to a broader view of education, one that includes a variety of policies. One way to think about this perspective is to use four Ps—purposes, policies, programs, and practices (Bybee 1997). Usually, individuals, organizations, and agencies contribute in various ways to the formulation of purposes, policies, programs, and practices; however, there must be coordination and consistency among the various efforts. Designing and developing new programs, as we did in the Sputnik era, surely marginalizes the success of the initiative if we do not attend to policies to support both those programs and changes in classroom practices to align with the innovative program.

Pay Attention to Equity.

Restricting initiatives to curriculum programs for specific groups of students (i.e., science and mathematically prone and college-bound students) resulted in criticism of Sputnik-era reforms as inappropriate for other students, such as average or disadvantaged students. To the degree that school systems implemented the new programs, teachers found that the materials were inappropriate for some populations of students and too difficult for others. Restricting policies or targeting programs opens the door to criticism on the grounds of equity. Ironically, proposing initiatives for *all* students also results in criticism for not addressing specific groups.

This lesson presents a major paradox of curriculum reform. To paraphrase Abraham Lincoln, you can please some of the people some of the time, but you will never please all of the people all of the time. My recommendation is to be clear about what you are doing, and do not try to fool some of the people by telling them your program was designed to do something for which it was not designed.

Examining the nature and lessons of Sputnik-era reforms, as well as those that came before and after, clearly demonstrated that education reforms differ. Although this may seem obvious, we have not always paid attention to some of the common themes and general lessons that may benefit the steady work of improving science, mathematics, and technology education. Stated succinctly, those general lessons are to use what we know about education change; include all the key players in the education community; align policies, programs, and practices with the stated purposes of education; work on improving education for all students; and attend to the support and continuous professional development of science teachers because teachers are the most essential resource in the system of science education. In the next section, I address new national aspirations and the vital importance of curriculum reform as a complement to other essential initiatives, such as common core standards and assessments.

In the late 1950s, the United States responded to the Sputnik challenge from the Soviet Union by accelerating, broadening, and deepening efforts to reform science and technology education. At that time, the national aspiration was to send a man to the Moon and have him return safely to Earth. We needed scientists and engineers to fulfill this goal, and the education community responded. Now the United States is being challenged again. The new national aspiration includes maintaining our economic competitiveness and sustaining global environments. Our contemporary response should be to heed lessons from the Sputnik era and must include reforms at the instructional core of science education, which includes the curriculum.

National Aspirations for the 21st Century
Sustaining Global Environments and Resources

Scientific literacy is essential to an individual's full participation in society. The understanding and abilities associated with scientific literacy empower citizens to make personal decisions and participate appropriately in the formulation of public policies that affect their lives. Assertions such as these provide a rationale for scientific literacy as the central purpose of science education. Too often, however, such a rationale lacks connections that answer questions concerning "personal decisions—*"Concerning what?"*; "Fully participate—*in what?"*; or "formulate policies—*relative to what?"* One could answer these questions using contexts that citizens confront daily—for example, personal health, natural resources, natural hazards, and information at the frontiers of science and technology. One other domain stands out—the environment. In the following discussion, I center on environmental issues as one context for reform of science curricula.

Environmental issues are a global concern. For more than a decade, climate change has been central to science and public policy at the local through global levels. Human activities such as the accumulation of waste, fragmentation or destruction of ecosystems, and depletion of resources have had a substantial effect on the global environment. As a result, threats to the environment are discussed prominently in the media, and citizens of every nation increasingly face the need to understand complex environmental issues. Noted scientist Edward O. Wilson summarizes the situation by using an economic metaphor:

What humanity is inflicting on itself and Earth is, to use a modern metaphor, the result of a mistake in capital investment. Having appropriated the planet's natural resources, we chose to annuitize them with a short-term maturity reached by progressively increasing payouts. At the time it seemed a wise decision. To many it still does. The result is rising per-capita production and consumption, markets awash in consumer goods and grain, and a surplus of optimistic economists. But there is a problem: the key elements of natural capital, Earth's arable land, ground water, forests, marine fisheries, and petroleum, are ultimately finite and not subject to proportionate capital growth. (2003, p. 149)

Wilson's use of an economic metaphor and my selection of this particular quotation are deeper and more insightful than they may seem. Citizens often hear economic arguments for the continued use of resources and destruction of environments. What Wilson's metaphor points out is the need to understand scientific ideas such as renewable and nonrenewable resources and ecosystems' capacity to degrade waste. Stated succinctly, understanding issues of ecological scarcity directly influences economic stability and social progress (Ophuls 1977). I maintain that ecological scarcity directly relates to environmental issues and a citizen's scientific literacy.

A scientifically literate individual has more than knowledge of environmental issues. A scientifically literate individual also must have attitudes that contribute to actions. Although not totally unrelated to civic attitudes and values, the attitudes referred to here are grounded more in an understanding of the environment and less in democratic values. Examples of values associated with the environment include conservation, prudence, and stewardship (Kollmuss and Agyeman 2002; Morrone, Manci, and Carr 2001; Tikka, Kuitunen, and Tynys 2000).

Asking and Answering the Sisyphean Question—Again

Here is one variation of the Sisyphean question in science education: What is important for citizens to know, value, and be able to do in situations involving natural resources and the environment?

For three decades, I have answered this question in a variety of forms and venues. My answers generally have been consistent, and the urgency of an explicit and direct response has only increased during the decades. So I see little need for a different statement, only the necessity for a coherent and sensible response by the larger science education community. The following response is for the most part a contemporary statement that is consistent with and builds on earlier recommendations (see, e.g., Bybee 1979a, 1979b, 1979c, 1984, 1991, 2003).

Being Clear About the Purposes of K–12 Science Education

Education has the purpose of preparing students for their future as citizens. As such, education should prepare students with the fundamental knowledge, skills, abilities, and sensibilities for the various situations they will fulfill in work and as citizens. During the K–12 years, education should center on students' general education and prepare them for both career and college. This view is a 21st-century perspective. In the past, students often were encouraged or counseled onto a college or vocational path. Now, the requirements for entering a career just out of high school or entering college are the same.

So how can one express the purpose of K–12 science education? The term *scientific literacy* expresses the general education goal described in the prior paragraphs. The PISA 2006 framework for science defines scientific literacy in terms of an individual:

- *Scientific knowledge and use of that knowledge to identify questions, to acquire new knowledge, to explain scientific phenomena, and to draw evidence-based conclusions about science-related issues.* These phrases express the central components of scientific literacy. For example, when individuals read about a health-related issue, can they separate scientific from nonscientific aspects of the text, and can they apply knowledge and justify personal decisions?
- *Understanding of the characteristic features of science as a form of human knowledge and enquiry.* For example, do individuals know the difference between evidence-based explanations and personal opinions?
- *Awareness of how science and technology shape our material, intellectual, and cultural environments.* This component of scientific literacy centers on the influence of science and technology on society. Can individuals recognize and explain the role of technologies as they influence a nation's economy, social organization, and culture? Are individuals aware of environmental changes and the results of those changes on economic and social stability?
- *Willingness to engage with science-related issues, and with the ideas of science, as a reflective citizen.* Finally, this dimension of scientific literacy underscores the attitudinal dynamics of scientific literacy. Are students interested in science? Memorizing and reproducing information does not necessarily mean students will select scientific careers, engage in science-related issues, or be willing as citizens to see public money allocated to scientific and technological research.

Establishing Policies for School Programs and Classroom Practices

Following is a discussion of education policies that are guidelines for science education programs, instruction, and practices. The policies are based on the fundamental divisions of ecology—organisms, environments, and populations. Using this ecological model and placing it in a human context, I asked, What is it about these divisions that is essential from a global perspective of sustainable development? My answers included both a conceptual and ethical orientation. Here are the answers, stated as policies. Science education programs and practices should guide learning toward (1) understanding and fulfilling basic human needs and facilitating personal development, (2) maintaining and improving the physical environment, (3) conserving natural resources and using them wisely, and (4) developing an understanding of interdependence between people at the local, national, and global levels, that is, development of a sense of community.

The ideas inherent in the first policy are simple and straightforward: All humans have basic physiological needs, such as clean air and water and sufficient food. They also need adequate shelter and safety. At higher levels, humans have the need to belong to groups and to perceive themselves as adequate and able. Simply stated, individuals need sustenance, order, community, and purpose for healthy physical, psychological, and social development. Education programs can contribute directly to the fulfillment of students' basic needs. They can be designed to help individuals gain knowledge about fulfilling these needs, inform individuals about the unfulfilled needs of others, and present the problems and possibilities associated with fulfilling human needs. The policy has a universal nature. All individuals have basic needs. Food and the development of a personal identity are both needs. Individuals in developed nations often think that alleviation of hunger and freedom from disease are the only basic needs in developing countries. The hierarchy of needs makes it clear that individuals in all nations are influenced by needs, though the needs may be different from one individual to the next and from one country to the next. A principal function of any society is to fulfill the needs of its citizens.

Science educators recognize only part of the problem, however, by presenting ideas that can help fulfill basic human needs. In *State of the World* (1990), Lester Brown and his colleagues clarify the role of values:

> *In the end, individual values are what drive social changes. Progress toward sustainability thus hinges on a collective deepening of our sense of responsibility to the earth and to future generations. Without a re-evaluation of our personal aspirations and motivations, we will never achieve an environmentally sound global community.* (Brown et al. 1990, p. 175)

To have any effect, policies must include both ideas and values, and it is essential that the values are compatible with the policy and serve to direct personal decisions toward achieving and maintaining sustainable growth. The values of justice and beneficence underlie the policy designed to fulfill basic human needs. With resource scarcity and a majority of world citizens with unfulfilled basic needs such as food, developed countries can no longer afford unnecessary goods and overconsumption, even for the cause of economic growth and the claims that all people are living a better life relative to the past.

Achieving this aim requires beneficence toward others, a value that can restrain personal consumption and encourage greater sharing. In turn, justice encourages the fair and equitable distribution of goods and services. This policy is more than an appeal to altruism. Adoption of green lifestyles that make use of appropriate goods and services in developed countries not only helps those in less developed countries but also better fulfills our own actual needs.

The second policy for programs and practices is designed to care for and improve the natural environment. Air, water, and soil are the common heritage of humankind, and they are essential to fulfilling basic needs. Many individuals perceive the environment as a receptacle of unlimited capacity to receive and degrade waste. But environmental systems are limited. The negative synergistic effects of pollution are becoming clear. Global warming and climate change are examples of this idea writ large. Realizing our dependence on the environment establishes a moral obligation to both ourselves and future generations to see that the environment can sustain life. Education programs should enable individuals to make informed decisions and take appropriate actions, in the short and long terms, to maintain and improve the physical environment.

The third policy concerning the conservation and wise use of resources is closely related to improvement of the physical environment and fulfillment of both the physical environment and basic needs. Just as we once believed in the limitless capacity of the environment to degrade waste, so too we once thought that resources were unlimited. They are not. Education about sustainable development will inform students of the need for resources, transitions to renewable resources, and the conservation of nonrenewable resources.

If one perceives the environment and resources as unlimited, then it is unnecessary to make value judgments about their use. The aim of sustainable development has an ecological ethic grounded in the idea of limited environmental capacities and limited depletion of resources. This, in a word, is prudence. Likewise, those with a vision of sustainability must think of themselves as stewards: managers and administrators of our natural environment.

The fourth and final policy is to develop increased interaction among people through education. This policy is directed toward establishing a greater sense of community. If fulfillment of human needs, improvement of the environment, and conservation of resources are to become realities, we must increase community involvement and cooperative participation at all levels, from local to global. One of the first steps toward productive personal interaction is the elimination of prejudicial barriers to community. Specifically, education programs should reduce prejudices such as racism, sexism, ethnocentrism, and nationalism. As long as one individual, group, or nation has a need to dominate another, the opportunities for harmonious living are reduced and the possibilities for disastrous conflict increase. Establishing a greater sense of community is clearly a prerequisite related to achieving the other three policies.

Cooperation and mutual regard are values essential for effective implementation of the fourth policy concerning growth and sustainable development. Inevitably, conflicts will arise among the crucial choices inherent in managing sustainable development. Societies can no longer afford to hold military force as the dominant means for resolving conflicts because force is ultimately divisive and results in destructive, not constructive, resolution of conflicts. Cooperative interaction is essential if all parties to a conflict are to achieve their goals and sustain a positive relationship. Finally, there is a profound need for a universal recognition of human rights and compassion for others. This is the value of mutual regard for each other now and consideration for future generations of humankind.

The education policies form a coordinated system of ideas and values supporting sustainable development. These policies would facilitate sustainable development while preserving personal freedom and minimizing governmental control. Education based on these policies could simultaneously produce changes in the ideas and values of individuals and implement means of regulating social change. Regulations, however, would not necessarily be the unilateral imposition of rules and laws by an authority on the majority. They would be, to use Garrett Hardin's phrase from his classic article "Tragedy of the Commons," "mutual cohesion mutually agreed upon" (Hardin 1968). Two factors justify this assertion. First, the ideas (needs, environment, resources, and community) and the values (justice, beneficence, stewardship, prudence, cooperation, and mutual regard) are sources of personal obligation as well as social regulation. Individuals with these ideas and values would be inclined to make informed decisions concerning their needs, the needs of others, the environment, and resources; practice self-restraint and self-reliance as necessary; and participate in the democratic development of rules based on the concept of sustainability. Second, a specific type of obligation is also inherent in the ideas and values. The obligation is reciprocal. The concern is not only for oneself but also for other people and their environments and resources.

Education programs that emphasize a sense of reciprocal obligation would develop an individual's sense of duty to others and the natural environment. Obligation alone can be engendered through social rules and laws. But this type of obligation is unilateral and can easily become little more than obedience to authority. This tendency is reduced, but not eliminated, through reciprocity among people who respect one another and their environments. Many individuals in social groups are reciprocally obligated to one another, so this idea is neither uncommon nor unachievable. Reciprocal obligations are grounded in empathizing with other people, coordinating efforts to solve problems, recognizing different points of view, balancing good and bad, and cooperating in the resolution of conflict. Humankind must take this direction if it is to avoid human ecological catastrophes and develop patterns of sustainable development.

So the education policies proposed here converge on the goal of sustainability and preservation of personal freedom through development of reciprocal obligation. The view presented here follows a course of least-restrictive regulation on the individual based on the possibility of changing personal ideas and values through education. In other words, regulations would increasingly influence the decisions of those individuals whose ideas and values are aligned with the old vision of unlimited industrial growth. An individual's freedom would be maintained to the degree that education achieves the described policies, thus developing personal ideas and values supporting sustainable growth. Education would create a dynamic interaction between self-restraint and social restriction, and that interaction would maximize personal freedom while achieving sustainable development.

Concluding Discussion

In the early years of the 21st century, the science education community must respond to an important challenge: helping citizens develop a greater knowledge and appreciation for resources and environmental issues. In an earlier section, I quoted E. O. Wilson, who used an economic metaphor in describing the environmental situation and his proposed solution. Today, the importance of understanding natural resources and the environment is even more important than it was last year, a decade ago, or 50 years ago. Being scientifically literate about resources and the environment is essential for all citizens, not only in the United States but in the global community as well.

A sound understanding of the dividends on the investment in scientific literacy accrues to all students in the form of enhanced learning and achievement. Science teachers, however, control the rate of interest and, therefore, the potential to increase the investment. The interest rates, and thus dividends, are largely determined by the degree to which the teaching includes challenging science content; increased curricular coherences; and greater congruence with personal, social, and global contexts. We must renew and double efforts to facilitate students' interdependence with nature and responsibility for sustaining a healthy and healing environment.

Teaching Science as Inquiry and Developing 21st-Century Skills

Contemporary national aspirations also include maintaining economic competitiveness. The economic theme is a relatively short-term goal, and for science education it implies preparation of a 21st-century workforce. For the science teacher, this aspiration translates to skills and abilities that can be developed within the theme—teaching science as inquiry.

This chapter directly relates to instruction and the need to reform instructional strategies, particularly those associated with scientific inquiry, so they enhance students' development of 21st-century workforce skills.

National Aspirations for the 21st Century: Maintaining Economic Competitiveness

In recent years, American businesses and industries have released numerous reports calling for major reform of our education system. Based on these calls for reform, for which *Rising Above the Gathering Storm* (National Academies 2005) must be considered the symbol of this national aspiration, one can conclude there is a significant need for education reform.

To sustain the U.S. position as a global competitor, our nation needs a first tactical response and eventually a strategic plan that outlines a decade of actions for reforming science and technology education. Although the need to change seems clear, the changes specifically implied for science and technology education for kindergarten through grade 12 are less than clear. This section advances concrete ideas that amount to a first tactical response, an opportunity that is available and generally supported by international, national, and state science education standards—teach science as inquiry.

Trends in Work Skills and Abilities

First, a brief examination of skills needed in the workplace sets the stage for implied changes in the science classroom. Figure 7.1 (p. 128) is a chart from labor economists and a perspective not often presented in science education discussions.

Figure 7.1

Trends in Routine and Nonroutine Task Input in the United States Since 1969

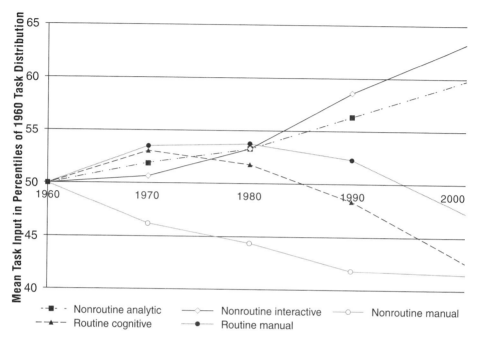

Nonroutine analytic   Nonroutine interactive   Nonroutine manual
Routine cognitive   Routine manual

Source: Autor, D., F. Levy, and R. J. Murnane. 2003. The skill content of recent technical change. *Quarterly Journal of Economics* 118 (4): 1279–1333.

Figure 7.1 shows a decline in tasks involving physical work that uses deductive or inductive rules. The figure also shows a decline in skills involving physical tasks that cannot be described as following a set of If-Then-Do rules. The latter has proven extremely difficult for computer programmers. This represents a decline in manual labor.

Much less attention has been devoted to the significant decline in *routine cognitive* task input, involving mental tasks that are well described by logical rules. Because such tasks can be accomplished by following a set of rules, they are prime candidates for computerization, and the figure shows, indeed, that demand for this task category has seen the steepest decline during the recent decade. Furthermore, rules-based tasks also are easier than other kinds of work to send to foreign producers. When a task can be reduced to rules—that is, a standard operating procedure—the process needs to be explained "only once," so the process of communicating with foreign producers is much simpler than the case of non-rules-based tasks where each piece of work is a special case. By the same token, when a process can be reduced to rules, it is much easier to monitor the quality of output. All of this highlights an important issue for school science: If students learn only to memorize and reproduce knowledge and skills, they risk being prepared only for jobs that are, in fact, increasingly disappearing from labor markets. In other words, the kind of learning outcomes that are emphasized in many school science programs and easiest to teach and assess are no longer sufficient to prepare students for the 21st-century.

In contrast, Figure 7.1 displays sharp increases in the demand for abstract tasks *requiring complex* communication, which involves interacting with humans to acquire information, explain it, or persuade others of its implications for action. Examples include a business manager motivating the people, a salesperson assessing a customer's reaction to a piece of equipment, a biology teacher explaining how cells divide, and an engineer describing a new design. Similar increases have occurred in the demand for *expert thinking*, which involves solving problems for which there are no rules-based solutions. Examples include diagnosing the illness of a patient with unique symptoms and repairing an auto that does not run well but for which the computer diagnostics report indicates no problem. These problems require what is referred to as pure pattern recognition—information processing that cannot yet be programmed on a computer. Although computers cannot substitute for humans in these tasks, computers can complement human skills by making information more readily available (Levy and Murnane 2004).

Connecting Science as Inquiry and 21st-Century Workforce Skills

The world economy has changed. Now, new skills will be required for the promising jobs and the 21st-century workforce. Economists have estimated that as much as half of the post–World War II growth in gross domestic product (GDP) in the United States is attributable to technological progress that resulted from research and development. In contrast, labor economists now warn that more than 50% of our children may leave school without the skills they need to enter the middle class. Business organizations produce reports such as *Building a Nation of Learners* and *Tapping America's Potential*, which suggest that many companies are having an increasingly difficult time finding employees with the critical-thinking, problem-solving, and communications skills they need to do their jobs. A rigorous education in science can help prepare students for good jobs, even if they never become scientists or engineers.

Most scientists would argue that science is an important tool for understanding the way the world works, comprehending some of the critical issues of the day, and even improving citizenship. The most compelling rationale for many parents might be to prepare their sons and daughters with skills they will need to prosper in a 21st-century workforce.

This perspective certainly does not mean to imply that everyone in the 21st century will be working as a scientist or engineer. Depending on how strictly we define science, only 3% to 8% of our workforce is employed as scientists, engineers, or other similar technical professionals. When we look at the skills valued by employers, however, it becomes apparent that science teachers have a lot to offer in preparing students to meet these proficiencies. To understand what this means, we will briefly discuss the skills valued by today's employers and examine how well students in the United States are learning these skills.

In the 1970s, students could achieve a middle-class lifestyle with only a high school diploma. Over the past 30 years, however, the skills needed to obtain a job and make a middle-class salary have changed dramatically. During this time, advances in technology, especially in manufacturing industries that had formerly paid high wages, along with the increasing international trade competition for low-skill jobs, have made things more difficult for students with only a high school diploma.

These difficulties are compounded by the fact that the skill set taught in schools has basically remained the same over this time period. The education that was effective in the 1970s has remained in place while the workplace has changed dramatically. In the early years of the 21st century, there is a dramatic gap between the abilities of graduating high school seniors and the skills valued by 21st-century employers.

In *Teaching the New Basic Skills: Principles for Educating Children to Thrive in a Changing Economy* (1996), Richard Murnane and Frank Levy defined a new set of skills important to employee-recruiting and work practices in firms paying high wages. The new basic skills are those abilities needed to obtain a middle-class position. They include

Hard skills
- the ability to read at the ninth-grade level
- the ability to do math at the ninth-grade level
- the ability to solve semistructured problems where hypotheses must be formed and tested

Soft skills
- the ability to work in groups with persons of various backgrounds
- the ability to communicate effectively, both orally and in writing

Other skills
- the ability to use personal computers to carry out simple tasks such as word processing

In today's advanced technological world, many employers are willing to teach knowledge specific to the industry as long as potential employees are proficient in these common abilities. In addition, these skills are needed by all students, regardless of whether they attend college or enter the workforce directly after high school. Recent assessments of 12th-grade students in the United States show that they lack preparation for these basic skills. According to findings from the 2006 National Assessment of Education Progress (NAEP), the number of students performing at a basic level in reading dropped by 7% between 1992 and 2005. In addition, just more than half of students performed at a basic level in mathematics. The Organization for Economic Cooperation and Development's (OECD) Programme for International Student Assessment (PISA) found that in 2003, 58% of students surveyed in the United States scored only as basic problem solvers, the lowest level of problem-solving ability. Furthermore, the United States ranked 29th out of 40 countries surveyed in this study (Lemke et al. 2005).

Inquiry shifts the focus of education to cognitive abilities such as reasoning with data, constructing an argument, and making a logically coherent explanation. On the most basic level, inquiry refers to the process of doing science. Inquiry-based learning engages students in the investigative nature of science. Using inquiry to teach science helps students put materials into context; fosters critical thinking; engages students more fully, resulting in positive attitudes toward science (Kyle et al. 1985; Rakow 1986); and improves communication skills (Rodriguez and Bethel 1983).

Inquiry can contribute to all students' education, not just those planning to attend college and major in science. Whether for a stock broker analyzing investment strategies, an auto mechanic identifying a problem, or an airline agent finding the best combination of seating, price, and schedule for a trip, effective reasoning and communication skills are vital.

Teaching Science as Inquiry in the 21st Century

This section synthesizes insights and ideas from prior sections and presents a contemporary view for the theme of teaching science as inquiry. By now, several points should be clear. First, teaching science as inquiry has a deep and rich intellectual foundation in American education. Second, there is an emerging foundation of research supporting the theme teaching science as inquiry, especially if one expands the idea from lessons to "integrated instructional sequences." Third, there is evidence, disappointing as it is, that teaching science as inquiry has never been fully realized in school science programs. The greatest efforts were during the Sputnik era, but they have not been sustained in contemporary school science programs and practices. In short, we have talked and written more about teaching science as inquiry than we have built and implemented programs that make real our goals and aspirations.

The Contemporary Challenges

To date, most discussions about the need to develop 21st-century workforce abilities and skills have centered on basic literacy and fundamental mathematics. To state the obvious, the potential of science education to be a major contributor to the development of a 21st-century workforce has been overlooked or ignored. Now is the time to remedy that situation.

The challenges we face can be summarized. Science education policies, programs, and practices should contribute to the development of students who have

- basic literacy (e.g., reading, writing, speaking)
- basic math (e.g., arithmetic, algebra, statistics)
- basic science competencies (e.g., identify scientific issues, explain phenomena scientifically, use scientific evidence)
- "hard" skills (e.g., problem solving, ability to apply science and mathematics in new situations)
- "soft" skills (e.g., work with people from other cultures, write and speak well, think in a multidisciplinary way, evaluate information critically, solve problems creatively)
- basic work skills (e.g., personal accountability, time and workload management)

This list may seem unique and unusual for science teachers. Almost immediately, one will note that this list does not emphasize science concepts and processes. However, I would argue they are included in scientific competencies, especially "explain phenomena scientifically."

The unusual nature of this list will engage the question, How can a science teacher possibly help students attain these basics? My response is that we have the opportunity before us if we teach science as inquiry.

Developing 21st-Century Skills in Science Classrooms

In 2007, the National Academies held two workshops that identified five broad skills that accommodated a range of jobs from low-skill, low-wage service to high-wage, high-skills professional work. Individuals can develop these broad skills within science classrooms as well as in other settings (NRC 2008, 2000; Levy and Murnane 2004). The skills identified, based on the National Academies workshops, are displayed in Figure 7.2.

A review of Figure 7.2 reveals a mixture of cognitive abilities, social skills, personal motivation, conceptual knowledge, and problem-solving competency. Although diverse, this knowledge and many of these skills and abilities can be developed in inquiry-oriented science classrooms. That said, it should be made clear that science teachers cannot, and probably should not, assume complete responsibility for developing all 21st-century skills. Even so, inquiry-oriented science classrooms have the opportunity to make a substantial contribution.

Figure 7.2

Examples of 21st-Century Skills

Research indicates that individuals learn and apply broad 21st-century skills within the context of specific bodies of knowledge (NRC 2008, 2000; Levy and Murnane 2004). At work, development of these skills is intertwined with development of technical job content knowledge. Similarly, in science education, students may develop cognitive skills while engaged in study of specific science topics and concepts.

1. **Adaptability:** The ability and willingness to cope with uncertain, new, and rapidly changing conditions on the job, including responding effectively to emergencies or crisis situations and learning new tasks, technologies, and procedures. Adaptability also includes handling work stress; adapting to different personalities, communication styles, and cultures; and physical adaptability to various indoor or outdoor work environments (Houston 2007; Pulakos et al. 2000).

2. **Complex communications/social skills:** Skills in processing and interpreting both verbal and nonverbal information from others to respond appropriately. A skilled communicator is able to select key pieces of a complex idea to express in words, sounds, and images to build shared understanding (Levy and Murnane 2004). Skilled communicators negotiate positive outcomes with customers, subordinates, and superiors through social perceptiveness, persuasion, negotiation, instructing, and service orientation (Peterson et al. 1999).

3. **Nonroutine problem solving:** A skilled problem solver uses expert thinking to examine a broad span of information, recognize patterns, and narrow the information to reach a diagnosis of the problem. Moving beyond diagnosis to a solution requires knowledge of how the information is linked conceptually and involves metacognition—the ability to reflect on whether a problem-solving strategy is working and to switch to another strategy if the current strategy isn't working (Levy and Murnane 2004). This ability includes creativity to generate new and innovative solutions, integrate seemingly unrelated information, and entertain possibilities others may miss (Houston 2007).

4. **Self-management/self-development:** Self-management skills include the ability to work remotely, in virtual teams; to work autonomously; and to be self-motivating and self-monitoring. One aspect of self-management is the willingness and ability to acquire new information and skills related to work (Houston 2007).

5. **Systems thinking:** The ability to understand how an entire system works, how an action, change, or malfunction in one part of the system affects the rest of the system; adopting a "big picture" perspective on work (Houston 2007). Systems thinking includes judgment and decision making; systems analysis; and systems evaluation as well as abstract reasoning about how the different elements of a work process interact (Peterson et al. 1999).

Clearly, statements of 21st-century skills must be adapted for science classrooms. This section describes strategies and contexts for the five skills (see Table 7.1 [pp. 136–137] for specific examples) in the context of school science programs. This discussion also presents connections to the *National Science Education Standards* (NRC 1996), in particular the Standards on inquiry.

Adaptability

Science programs will provide learners with experiences that require coping with new approaches to investigations, analyzing less-than-clear data, using new tools and techniques to make observations, and collecting and analyzing data. Programs will include opportunities to work individually and in groups on science activities, investigations, laboratories, and field studies.

Specific examples include the following:

- Use appropriate tools and equipment to gather, analyze, and interpret data.
- Design and conduct a scientific investigation.

Complex Communications/Social Skills

Programs with varied learning experiences, including laboratories and investigations, will require students to process and interpret information and data from a variety of sources. Learners would have to select appropriate evidence and use it to communicate a scientific explanation. Science programs would include group work that culminates with the use of evidence to formulate a conclusion or recommendation.

Specific examples include the following:

- Design and conduct scientific investigations (with a group).
- Communicate scientific procedures and explanations, as well as defend a scientific argument.
- Use technology and mathematics to improve investigations and communications.

Nonroutine Problem Solving

Science programs will require learners to apply knowledge to scientific questions and technological problems, identify the scientific components of a contemporary issue, and use reasoning to link evidence to an explanation. In the process of scientific investigations, learners will be required to reflect on the adequacy of an answer to a scientific question or a technological solution to a problem. Students may be required to think of another investigation or another way to gather data and connect those data with the extant body of scientific knowledge.

Specific examples include the following:

- Identify questions that can be answered through scientific investigations.
- Develop descriptions, explanations, predictions, and models using evidence.
- Think critically and logically to make the relationship between evidence and explanations.
- Recognize and analyze alternative explanations and predictions.

Self-Management/Self-Development

Programs will include opportunities for students to work on scientific investigations alone and as a group. These investigations would include full inquiries and may require learners to acquire new knowledge and develop new skills as they pursue answers to questions or solutions to problems.

Specific examples include the following:

- Design and conduct a scientific investigation.
- Use appropriate tools and techniques to gather, analyze, and interpret data.

Systems Thinking

School science programs would include the introduction and applications of systems thinking in the context of life, Earth, and physical science as well as multidisciplinary problems in personal and social perspectives. Learners would be required to realize the limits to investigations of systems; describe components, flow of resources, and changes in systems and subsystems; and reason about interactions at the interface between systems.

Specific examples include:

- Identify questions that can be answered through scientific investigations.
- Design and conduct a scientific investigation.
- Think critically and logically to make the relationship between evidence and explanation.

Table 7.1 (pp. 136–137) summarizes essential features of the skills and provides examples for school science programs.

Concluding Discussion

Addressing the need to develop 21st-century workforce skills will require students to have experience with activities, investigations, and experiments. In a word, science teaching needs to be inquiry-oriented. This orientation seems obvious, but it must be emphasized. Science education has an opportunity to make a substantial contribution to one of society's pressing problems. Science classrooms provide the setting for helping students learn most, if not all, of the skills described in Table 7.1 (pp. 136–137). To accomplish this goal, science educators must provide opportunities for students to adapt to others' work styles and ideas, solve problems, manage their work, think in terms of systems, and communicate their results.

Table 7.1

Developing 21st-Century Skills in Science Programs

Essential Features of 21st-Century Skills	Examples of Contexts for School Science Programs
Adaptability	
• Cope with changing conditions • Learn new techniques and procedures • Adapt to different personalities and communication styles • Adapt to different working environments	• Work on different investigations and experiments • Work on investigations or experiments • Work cooperatively in groups • Work on investigations in the laboratory and outdoors
Complex Communications/Social Skills	
• Process and interpret verbal/nonverbal information • Select key pieces of complex ideas to communicate • Build shared understanding • Negotiate positive outcomes	• Prepare oral and written reports communicating procedures, evidence, and explanations of investigations and experiments • Use evidence gained in investigations as a basis for a scientific explanation • Prepare a scientific argument • Work with group members to prepare a report
Nonroutine Problem Solving	
• Use expert thinking in problem solving • Recognize patterns • Link information • Integrate information • Reflect on adequacy of solutions • Maintain several possible solutions • Propose new strategies • Generate innovative solutions	• Recognize the need to search for expert's knowledge • Recognize patterns in data • Connect evidence and information from an investigation with scientific knowledge from textbooks, the web, or other sources • Understand constraints in proposed solutions • Propose several possible solutions and strategies to attain the solutions • Propose creative solutions
Self-Management/Self-Development	
• Work remotely (individually) • Work in virtual teams • Self-motivate • Self-monitor • Have willingness and ability to acquire new information and skills	• Work individually at home • Work with a virtual group • Complete a full/open investigation • Reflect on adequacy of progress, solutions, explanations • Acquire new information and skills in the process of problem solving and working on investigation

Table 7.1 *(continued)*

Developing 21st-Century Skills in Science Programs

Essential Features of 21st-Century Skills	Examples of Contexts for School Science Programs
Systems Thinking	
• Understand the systems concept • Understand how changes in one part of the system affects the system • Adapt a "big picture" perspective • Systems analysis • Judgment and decision making • Abstract reasoning about interactions among components of a system	• Describe components of a system based on a system under investigation • Predict changes in an investigation • Analyze a system under investigation • Make decisions about best proposed solutions • Demonstrate understanding about components and functions of a proposed system

Learning outcomes aligned with inquiry and 21st-century skills can be attained using both full and partial inquiries. Central to these skills are group work and cognitive abilities such as reasoning. Although some may argue for full inquiries, and I agree that these should be part of a student's science experience, there is a place for partial inquiries. After all, the emphasis is on the learning outcomes, and these may be achieved with partial inquiry experiences. The important point is to give emphasis to the skills and abilities described earlier.

One challenge for curriculum, instruction, and assessment is implementing what I have called *integrated instructional sequences.* A National Research Council report, *America's Lab Report: Investigations in High School Science* (Singer, Hilton, and Schweingruber 2006), introduced the idea as "Integrated instructional units connect laboratory experiences with other types of science learning activities, including lectures, reading, and discussion" (p. 4). The BSCS 5E Instructional Model is one example of an integrated instructional unit. In a paper prepared for a National Research Council workshop exploring the intersection of science education and the development of 21st-century skills, I described the research supporting the BSCS 5E model and potential linkages with 21st-century skills (Bybee 2009).

Using the BSCS 5E Instructional Model or another variation on the learning cycle provides connections among curriculum, instruction, and assessment and enhances students' opportunities to attain learning outcomes, including 21st-century skills.

This chapter has provided clarification of 21st-century skills in the context of science education programs and practices. Here are some concrete recommendations that science education leaders can use as they implement changes that will promote 21st-century skills as learning outcomes.

Make sure all students meet the standards for scientific inquiry and technological design. Beginning with the national standards and extending to state and local standards, abilities related to scientific inquiry are included as learning outcomes. Statements of the need to develop the abilities of scientific inquiry and technological design can be the connection between what many will perceive as the abstract vision of 21st-century skills and the concrete context of science teaching.

Build on the opportunities that already exist in school programs and teaching practices. Understandably, many will see the call for development of 21st-century skills as a major change, one beyond their capabilities and interests. Centering the changes on opportunities that already exist in investigations, laboratories, and activities will soften the resistance to change. In many cases, science teachers already contribute to the development of these skills; the change is one of clarity and emphasis. In particular, one of the changes that may be new for science teachers includes placing an emphasis on individual and interpersonal skills.

Emphasize cognitive abilities and skills as learning outcomes. Bringing the development of cognitive abilities and interpersonal skills to the foreground in the science classroom may be new to science teachers. Providing teachers with statements they can use, such as "What is the evidence for that explanation?" "What alternative explanations have you heard from the team?" "What goals of the investigation include working together to gather evidence and form an explanation?" will help.

Use the idea of integrated instructional sequences. Helping science teachers connect lessons will provide the time and opportunity needed for the emphasis on 21st-century skills. In addition, using the sequences will enhance the opportunities for other learning outcomes. Of course, I recommend using the BSCS 5E Instructional Model, but the important idea is to use an integrated instructional sequence, not one particular model.

Include basic skills of literacy and mathematics as part of learning outcomes. Because part of the student's work will include the presentation of results, graphs, charts, diagrams, and reports, the inclusion of basic literacy and mathematics should be considered part of a new emphasis on 21st-century skills.

The evidence indicates that teaching science as inquiry is not now, and never has been, in any significant way, implemented. Furthermore, it probably would not matter much what one used as a definition—inquiry as science content or inquiry as teaching strategies. Probably the closest the science education community came to teaching science as inquiry was during the 1960s and 1970s as we implemented the Sputnik-spurred curriculum programs and provided massive professional development experiences for teachers. The evidence does indicate that these programs were effective for the inquiry-oriented objectives that were emphasized in that era. Although science educators continue to state the need for and importance of teaching science as inquiry, our state policies, assessments, textbooks, and classroom practices give little evidence of the realization of this goal.

In this chapter, I have restated and provided details of what we mean by teaching science as inquiry. Appropriately viewed, inquiry as science content and inquiry as teaching strategies are two sides of a single coin. Teaching science as inquiry means providing students with diverse opportunities to develop the abilities and understandings of scientific inquiry while also learning the fundamental subject matter of science. The teaching strategies that provide students those opportunities are found in varied activities, laboratory investigations, internet use, and student-initiated inquiries, all presented in integrated instructional sequence. Science teachers know this simple educational insight. Now we are called to the new challenges of the 21st century. We have the opportunity to show that the science education community will respond to a great national need by teaching science as inquiry.

A Perspective on the Reform of Science Teaching

After the launch of Sputnik in October 1957, the United States responded to the Soviet Union by accelerating, broadening, and deepening efforts to reform science and technology education. Now our country is being challenged again. Our contemporary response again must include improving science education in general and, relative to themes in this book, science teaching in particular.

The U.S. response to Sputnik was unique to that time in history. So, too, must the contemporary response be unique. Now the primary goals are to sustain innovation by both scientists and engineers, create a deep technical workforce, and develop scientifically and technologically literate citizens for the 21st century. All of us—science teachers, teacher educators, policy makers, and the public— must ask and answer the Sisyphean question: What should citizens know, value, and be able to do in preparation for life and work in the 21st century?

This chapter presents a perspective on reform. It begins with a brief review of the instructional core, then turns to a larger view of reform, one that includes broader questions of goals and progresses to the most fundamental area—classroom practices. After this overview of reform, I address practical questions of what must be done to improve science teaching and respond to the 21st-century goals—scientific literacy, a deep technical workforce, and a diverse scientific and engineering workforce.

Stay Focused on the Instructional Core

What is meant by *instructional core*? In the simplest form, the instructional core consists of the students, teacher, and learning outcomes. Of course, the learning process becomes more complex when you consider the backgrounds and diversity of students in any classroom, qualifications of the teachers, and the difficulty of learning conceptual ideas and the complex processes of scientific inquiry. Richard Elmore (2009) has pointed out that there are only three ways to improve student learning at a scale that makes a difference. First, you can increase the rigor and focus of content. Second, you can increase the level of students' learning of content. Third, you can increase teachers' knowledge and skill for teaching the content (see Figure 8.1).

Figure 8.1

The Instructional Core

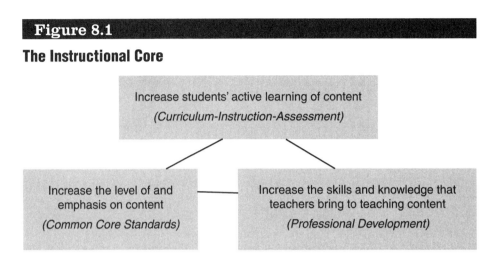

Source: Elmore, R. 2009. Improving the instructional core. In *Instructional rounds in education: A network approach to improving teaching and learning*, ed. E. U. City, R. Elmore, S. Fiarman, and L. Teite. Cambridge, MA: Harvard Education Press.

Changing the Content

Increasing the level or focus of content is usually the goal of revising national, state, or local content or performance standards. The content standards may, for example, aim to change science content from facts to major conceptual ideas and core concepts in science. The focus might change from an exclusive emphasis on scientific knowledge to a balance of scientific knowledge and scientific literacy. Whatever the change in content, decisions about the content and performance standards are controlled by groups and processes such as national organizations, state committees, or local teams.

Engaging the Learner

In most classrooms, changing the level of student learning is influenced by the school or district curriculum, instructional materials, and the strategies and methods of science teaching that teachers use. Instructional materials may facilitate teachers' understanding and use of strategies that change the level of engagement and learning through the introductions of strategies based on contemporary understanding of how students learn science. The BSCS 5E Instructional Model serves as an example of an integrated instructional sequence that gives teachers and students time and opportunities to learn new, challenging science content and develop abilities for innovation. The design of instructional materials can help teachers understand and apply strategies that will engage students, but the participating teachers have to make changes to accommodate their unique schools, courses, and students.

Providing Professional Development

The third type of change in the instructional core is a unique and most powerful contribution to improving student learning at scale. Increasing the teacher's knowledge of students' learning, their inquiry-based teaching skills, and instructional model use can provide the basis for engaging students actively in learning. Professional development can use a unique, constructive, and opportunistic approach to the instructional core, which has the potential to influence student learning at a scale that eventually will be evident in assessments.

Changing One Element Requires Changes in Two Others

However, there is, as Shakespeare pointed out, a "rub." Increasing one of the three essential elements of the instructional core requires changes in the other two. The National Research Council (NRC), National Governors Association (NGA), and Council of Chief State School Officers (CCSSO) are developing new standards for science education. So, increasing teachers' knowledge and skills requires some understanding of those standards and the subsequent need to change and reform curricula to enhance student engagement. This fact suggests the need to help teachers and administrators recognize the required changes in school programs if they want to increase student achievement at scale.

To conclude, the education landscape is littered with strategies, projects, models, materials, and innovations that respond to continuing calls for reform and improvement of student learning. Let's stop and ask, What really counts for improvement? The answer is student achievement. Whether determined by a traditional end-of-course grade, state tests, the national report card, or international assessments, student achievement is the bottom line. So, one can ask, What can educators do to improve student achievement? A second fundamental question follows: What can we do to improve student achievement at a scale that makes a difference? The answer is clear and direct: Stay focused on the instructional core.

Understanding the Dimensions and Dynamics of Science Education
The Purpose of Science Education

The term *purpose* refers to various goal statements of what science teaching should achieve, such as scientific literacy for all learners. The strength of purpose statements lies in their widespread acceptance and agreement among science educators and their application to all components of science education—for example, classroom teaching, teacher education, curriculum development, and policy making at local, state, and national levels. Weaknesses of purpose statements exist in their ambiguity about the role of specific components of science education. For example, what does the purpose of achieving scientific literacy mean for an elementary grade teacher? A high school Earth science teacher? A science supervisor? A curriculum developer? A teacher educator? Clearly, the answers vary. Individuals need statements representing scientific literacy that are more concrete and directly related to various components of science education.

National statements about the purposes of science education support the vision that science education must accommodate all students. Specifically, national standards define the level of understanding and the abilities that all students—regardless of background, future aspirations, or interest in science—should develop. By their position as national standards, these policy documents embody the assumption that all students can learn science, or, to paraphrase an aphorism from an earlier era of reform, science can be taught effectively in some intellectually honest form to all students (Bruner 1960).

National standards encourage science teachers to provide opportunities for all students to learn science throughout their school years. They clearly and unequivocally advocate including those who traditionally have not received encouragement and opportunities to learn science.

Policies for Science Education

Policy statements are concrete translations of the purpose—achieving scientific literacy for all learners—for various components of science education. Documents that give direction and guidance but are not actual programs serve this purpose. Examples of policy documents include district syllabi for K–12 science; state frameworks; and national, state, and local standards. In the contemporary reform movement, several documents clarify policies for scientific literacy. *National Science Education Standards* (NRC 1996), *Benchmarks for Science Literacy* (AAAS 1993), and the *Science Framework for the 2009 National Assessment of Educational Progress* (NAGB 2009)—all of which have considerable overlap and consistency for the content—provide clear, detailed, and elaborate definitions of scientific literacy. They represent common ground for the content of science education (AAAS 1995). Science teachers should expect the new "common core" standards for science to build on and complement current standards.

Concerning the dimensions of scientific literacy, the Standards and Benchmarks present a balance of functional, conceptual, procedural, and multidimensional scientific literacy. They have, for example, reduced technical words and thus represent a significant first step toward less emphasis on scientific vocabulary and more emphasis on other dimensions of scientific literacy. The documents elaborate on conceptual and procedural dimensions of scientific literacy. Furthermore, the Standards include changes from prevalent views of scientific processes. The abilities of inquiry, for instance, extend beyond a limited emphasis on science processes, such as observation, inference, hypothesis, and experiment. The Standards on "Science as Inquiry" include the processes of science and give greater emphasis to cognitive abilities, such as using logic, evidence, and extant knowledge to construct explanations of natural phenomena. Finally, the policy documents include the human dimensions of science and technology, such as history, the nature of science, and science in personal and social perspectives.

Programs for Science Teaching

Science programs include the actual curriculum materials based on policy documents such as the Standards and Benchmarks. Science programs are unique to grade levels, disciplines, and aspects of science teaching and present a consistent, coordinated, and coherent approach to the science education of all students. Examples of science programs for secondary schools include the American Chemical Society's *ChemCom* and the new Biological Sciences Curriculum Study's *BSCS Science: An Inquiry Approach*.

School science programs may be developed by national organizations, or they may be developed by states or local school districts. Who develops the materials is not the defining characteristic of science programs. That schools, colleges, state agencies, and national organizations have programs aligned with national, state, and local policies is the important feature and requirement of standards-based reform in the 21st century.

Practices for the Teaching of Science

Practice refers to the specific processes of teaching science. The practices of science teaching include the personal dynamics between teachers and students and the interactions among students and assessments, educational technologies, laboratories, and myriad other science teaching strategies. The view of contemporary reform described here assumes that science teachers will implement classroom practices consistent with policies, programs, and the goal of achieving scientific literacy for all learners. Improving the practices in the classroom centers on the instructional core and the most individual, unique, and fundamental aspect of science education—the act of teaching students. From the perspective of science teachers, there should be little doubt about the need for local leadership and support for their work in contemporary reform.

Dynamics of Contemporary Reform

If achieving scientific literacy is the goal and science teachers understand the various domains and dimensions of scientific literacy, then it seems important to have a map of the reform territory to know your location, means of movement, direction of travel, and what lies ahead. We can use themes just outlined—purpose, policy, program, practice—for locating and clarifying different efforts in the geography of contemporary reform (see Tables 8.1 [pp. 146–147] and 8.2 [p. 148]).

Table 8.1

Dimensions of Contemporary Reform

Perspectives	Time (for actual change to occur)	Scale (number of individuals involved)	Space (scope and location of the change activity)	
Purpose Reforming goals Establishing priorities for goals Providing justification for goals	*1–2 years* To publish document	*Hundreds* Philosophers and educators who write about aims and goals of education	*National/Global* Publications and reports are disseminated widely	
Policy Establishing design criteria for programs Identifying criteria for instruction Developing frameworks for curriculum and instruction	*3–4 years* To develop frameworks and legislation	*Thousands* Policy analysts, legislators, supervisors, and reviewers	*National/State* Policies focus on specific areas	
Program Developing materials or adopting a program Implementing the program	*3–6 years* To develop a complete educational program	*Tens of Thousands* Developers, field-test teachers, students, textbook publishers, software developers	*Local/School* Adoption committees	
Practices Changing teaching strategies Adapting materials to unique needs of schools and students	*7–10 years* To complete implementation and staff development	*Millions* School personnel, public	*Classrooms* Individual teachers	

The left column in Table 8.1 summarizes the perspectives of purpose, policy, program, and practices. The top row includes six aspects of reform: time, scale, space, duration, materials, and agreement. You can review the table and develop a general sense of the dimensions and difficulties of the reform effort as you ask questions such as the following:

- How long does it take to form policies such as national standards or state frameworks?
- Once a new program is implemented, how long will it continue in a school system?

Duration (once change has occurred)	Materials (actual products of the activity)	Agreement (difficulty reaching agreement among participants)
Year New problems emerge and new goals and priorities are proposed	*Articles/Reports* Relatively short publications, reports, and articles	*Easy* Small number of reviewers and referees
Several Years Once in place, policies are not easily changed	*Book/Monograph* Longer statements of rationale, content, and other aspects of reform	*Difficult* Political negotiations, trade-offs, and revisions
Decades Once developed or adopted, programs last for extended periods	*Books/Courseware* Usually several books for students and teachers	*Very Difficult* Many factions, barriers, and requirements
Several Decades Individual teaching practices often last a professional lifetime.	*Complete System* Books plus materials, equipment, and support	*Extraordinarily Difficult* Unique needs, practices, and beliefs of individuals, schools, and communities

- Who is responsible for a particular effort, such as curriculum reform, policy formation, or classroom practices?
- How do all dimensions of the framework contribute to the whole of science education?
- How does the framework relate to systemic initiatives?

Table 8.2 (p. 148) describes other aspects of reform. Again, the left column includes the perspectives of purpose, policy, program, and practices. The top row includes risk, cost, constraints, responsibilities, and benefits and considers these in terms of school districts, school personnel, and students. The analysis presented in the figure indicates that purpose statements and policy documents, although essential, have minimal and moderate influence on reform, respectively. We are now approaching the phases where risk, cost, constraints, personal responsibilities, and benefits are all high or extremely high. Clearly, the science teaching community has significant challenges ahead.

Table 8.2

Cost, Risks, and Benefits of Contemporary Reform

Perspectives	Risk to Individual School Personnel	Cost in Financial Terms to School	Constraints Against Reform for School	Responsibility for Reform by School Personnel	Benefits to School Personnel and Students
Purpose Reforming goals Establishing priorities for goals	Minimal	Minimal	Minimal	Minimal	Minimal
Policy Establishing design criteria Identifying criteria for instruction Developing frameworks for curriculum and instruction	Moderate	Moderate	Moderate	Moderate	Moderate
Program Developing materials or adopting a program Implementing the program	High	High	High	High	High
Practices Changing teaching strategies Adapting materials to unique needs of schools and students	Extremely high	Extremely high	Extremely high	Extremely high	Extremely high

Perhaps more important than the specific cells, Tables 8.1 and 8.2 give an overall picture of the reform effort. If I placed a "you are here" label on this map, it would be the interface between policy and program. We have policies in the form of the Standards and Benchmarks. The next phases of reform will take longer; involve more individuals, materials, and equipment; move closer to schools and classrooms; and present more difficulties when it comes to reaching agreement and actually improving school science programs and changing instructional practices.

The nation needs a vision, a first tactical response, and a strategic plan for a decade of actions, all designed to reform science education to develop scientific literacy and sustain the U.S. position as a global leader. Although the need to change seems evident, the changes specifically implied for science and technology for kindergarten through grade 12 must be clarified and addressed. The next sections are based on an article titled "Do We Need Another Sputnik?" (Bybee 2007) and a report titled *A Decade of Action: Sustaining Global Competitiveness* (BSCS 2007).

Fostering Scientific Literacy
What Must We Do?

I begin with a recommendation that will facilitate reform by beginning with teachers and their standard request when asked to change: Where are the materials?

Develop a new generation of curriculum materials for scientific literacy. Specifications for the curriculum materials use the contexts and competencies from PISA 2006 Science, and the content builds on both national and international frameworks. Based on PISA 2006, Figure 8.2 (p. 150) presents a framework for the curriculum. Contexts for the curriculum are described in Table 8.3 (p. 151).

Content for the curriculum would be based on the *National Science Education Standards* (NRC 1996) and the *Benchmarks for Scientific Literacy* (AAAS 1993), and aligned with the *Science Assessment and Item Specifications for the 2009 National Assessment of Educational Progress* (National Assessment Governing Board 2005) and the new common core standards.

Support professional development of science teachers. Specific actions are recommended to achieve this goal. First, establish summer institutes that focus on building teachers' content and pedagogical knowledge and skills. There should be follow-up experiences during the academic year. Second, develop online communities to support all participating science teachers. These professional development programs should be concentrated and continuous, have an educational context, focus on content, and establish professional learning communities.

Figure 8.2

A Perspective for K–12 Scientific and Technological Literacy

contexts

Life and work situations that involve
science and technology
require individuals to

competencies

- identify scientific and technological issues,
- acquire and use scientific and technical information,
- understand complex systems,
- use a variety of technologies, and
- apply thinking skills.

An individual's competence
is influenced by content and
abilities, skills, and attributes

content

- knowledge about the natural and
  designed world
- knowledge about science and technology
  as a domain of human endeavor

abilities, skills,
and attributes

- workplace competence
- foundational skills
  - basic skills
  - thinking skills
  - personal qualities

Table 8.3

Contexts for the Science Curriculum

Context	Personal	Social	Global
Careers	Scientific research, engineering, technical, and teaching	Scientific research, medicine, engineering, information and communication technology	World health, economic progress, security
Health	Maintenance of health, accident prevention, nutrition	Control of disease and social transmission, nutrition, food choices, community health	Epidemics and spread of infectious diseases
Resources	Control of personal consumption of materials and energy	Maintenance of human populations, quality of life, security, production and distribution of food, energy supply	Renewable and nonrenewable energy, natural systems, population growth, sustainable use of species
Environment	Research on environmentally friendly behavior, use and disposal of materials	Research on population distribution, disposal of waste, environmental impact, local weather	Biodiversity, ecological sustainability, control of pollution, production and loss of soil
Hazards	Natural and human-induced hazards, decisions about housing	Rapid changes (earthquakes, severe weather), slow and progressive changes (coastal erosion, sedimentation), risk assessment	Climate change, impact of modern warfare
Research and Development	Interest in science and technology, science-based hobbies, sport and leisure activities, use of personal technology	Aerospace engineering, biotechnology, information and communications technology, pharmaceuticals	Exploration of space, transportation, agriculture, applications to resolve global problems

The professional development programs should provide enough initial time to establish a clear foundation for teaching and learning. In addition to an early concentration, the program should extend over a year (or more) and include continuous work on selecting curriculum materials and improving instruction. The educational context for the professional development programs should include curriculum—that is, content and pedagogy with a direct and purposeful meaning for science teachers. Core concepts for scientific literacy must be the programs' focus. Finally, the programs require the establishment of professional learning communities, with teams of teachers analyzing teaching, engaging in lesson study, reviewing content, and working on the implementation of curriculum materials.

Align certification and accreditation with contemporary priorities of scientific literacy. This recommendation uses the critical leverage of science teacher certification to facilitate reform of undergraduate teacher education programs. No discussion of improving science education escapes acknowledging the need to change teacher education. This includes changes in states' certification and national accreditation. In addition, federal support to colleges and universities that prepare significant numbers of future science teachers will be a major contribution to their reform. To this recommendation I add special support to colleges and universities with significant populations of Hispanic, African American, and Native American students so the institutions can recruit and prepare a greater diversity of science teachers.

Build district-level capacity for continuous improvement of programs for scientific literacy. Specific actions necessary for this priority include developing leaders, providing summer programs and assistance during the year, centering on critical leverage points such as selection of instructional materials, and designing programs so the district builds a sustainable infrastructure.

This priority connects to other priorities with the goal of sustaining the initial results attained through professional development, curriculum reform, and reform of undergraduate education. Although the federal costs will be high initially, by building district-level capacity one could anticipate reduced support in the long-term.

Explain to the public what this school science reform is about and why it will benefit their children and the country. One of the great insights from the Sputnik era was the fact that national leaders provided clear and compelling explanations of what the reform was and why it was important. Furthermore, there was continued support for science teachers and a national enthusiasm for reform.

A Vision and a Plan

As I have tried to make clear, current national aspirations center on economic and environmental rationales for education reform. Stated succinctly, the rationales state that our economic security depends on educating people for life and work in the 21st century. For the most part, the science education community has not made general connections to the economic rationale. Furthermore, there have been reports but no reform initiatives that represent a positive, constructive response to demands for an improved workforce and greater scientific literacy.

The vision for this reform centers on *content* aligned with science education policies such as the 2009 National Assessment of Educational Progress and frameworks for the international assessments PISA and TIMSS. The *contexts* for science programs range from personal to global and include categories such as careers, health, resources, environment, hazards, and research and development.

The *competencies* important for 21st-century science literacy build on the Secretary's Commission on Achieving Necessary Skills (SCANS) and specifically emphasize those skills and abilities that may be developed in school programs. Although numerous reports from business, industry, and government are not explicit about skills for the 21st century, recent workshops conducted by the National Research Council have described a set of 21st-century skills. Figure 8.3 presents a framework that includes the key features of these 21st-century skills.

Figure 8.3

Examples of 21st-Century Skills

Development of the following skills is intertwined with development of content knowledge related to technical jobs. Similarly, in science education, students may develop cognitive skills while engaged in study of specific science topics and concepts.

Adaptability: The ability and willingness to cope with uncertain, new, and rapidly changing conditions on the job, including responding effectively to emergencies or crisis situations and learning new tasks, technologies, and procedures. Adaptability also includes handling work stress; adapting to different personalities, communication styles, and cultures; and adapting physically to various indoor or outdoor work environments.

Complex communications and social skills: Skills in processing and interpreting both verbal and nonverbal information from others to respond appropriately. A skilled communicator is able to select key pieces of a complex idea to express in words and images to build shared understanding. Skilled communicators achieve positive outcomes with customers, subordinates, and superiors through social perceptiveness, persuasion, negotiation, instruction, and a personal orientation.

Nonroutine problem solving: A skilled problem solver uses expert thinking to examine a broad span of information, recognize patterns, and narrow the information to reach a diagnosis of the problem. Moving beyond diagnosis to a solution requires knowledge of how the information is linked conceptually and involves the ability to reflect on whether a problem-solving strategy is working and to switch to another strategy if the current strategy isn't working. Problem solving includes creativity to generate innovative solutions, integrate seemingly unrelated information, and entertain possibilities others may miss.

Self-management/self-development: Self-management skills include the ability to work remotely, in virtual teams; to work autonomously; and to be self-motivating and self-monitoring. One aspect of self-management involves the willingness and ability to acquire new information and skills related to work.

Systems thinking: The ability to understand how an entire system works, how an action, change, or malfunction in one part of the system affects the rest of the system—adopting a "big picture" perspective on work. Systems thinking includes judgment and decision making; systems analysis; and systems evaluation as well as abstract reasoning about how the different elements of a work process interact.

Competencies are in a central position as they represent the essential change in emphasis for curricular supplements and teaching strategies described in the next sections. These skills have been mentioned in prior chapters. They are summarized here as basic to proposed instructional materials.

How We Can Begin

This section presents a larger picture of how we can initiate and bring about the changes described in the last section to a scale that matters within the U.S. education system.

The science education community must plan a decade of action. Achieving higher levels of scientific literacy cannot be accomplished quickly; it will take a minimum of 10 years. Tables 8.4 and 8.5 present specifications for reform and phases for a decade of reform centering on improving scientific literacy in the United States.

Table 8.4

Specifications for Action

Unit of Change	Instructional Core
Time frame for change	10 years
Critical core of change	Teachers' knowledge and skills, curriculum for active learning, level of content and abilities
Components of change	Education policies, curriculum programs, teaching practices
Theory of action for change	Introduce curriculum model instructional units for reform and provide professional development based on those units. Changes in assessment would be introduced as complements to curriculum reform.

Table 8.5

A Decade of Action: Phases and Goals

Phase	Timeline	Goal
Initiating a response	2 years	Design, develop, and implement model instructional units.
Bringing the reform to scale	6 years	Change policies, programs, and practices at local, state, and national levels.
Sustaining the reform	2 years	Build capacity at the local level for continuous improvement of school science and technology programs.
Evaluating the reform	Continuous, with major evaluation in 10 years	Provide formative and summative data on the nature and results of the reform efforts.

A Decade of Action

This section presents a strategic plan for making the vision a reality. The plan will require a *Decade of Action*. I use *The Tipping Point* (Gladwell 2002) as the theory of action and identify school districts as the unit of change.

Initiating the Reform: Introducing Little Changes With Big Effects

The work for this phase will last two years. Beginning with a brief period of dialogue to form partnerships and establish coalitions of support, this phase very quickly turns to the funding and development of *model instructional units* for reform. The model instructional units use major sectors of the economy as the "topics" (e.g., aerospace, biotechnology, energy, hazard mitigation, health, and environmental quality) and emphasize themes such as careers and research and development.

Providing model instructional units, professional development, and exemplary assessments at the elementary, middle, and high school levels will have an effect on the system, develop understanding among school personnel, and increase support by policy makers and administrators. Furthermore, the units will provide a basis for answering the public's questions about what the changes involve and why they are important—especially for students.

Bringing the Reform to Scale: Systematic Changes That Make a Difference

Bringing the reform to scale will take six years. During this time, some work will continue on instructional materials developed in the first phase. After the initial phase, efforts to bring the reform to a significant scale would expand. Evaluations of teachers' responses and students' achievements, abilities, and attitudes would be reviewed and analyzed. These data would form the basis for revision of the original modules, development of new modules, and a compelling case statement for continued expansion of the reform. This is when the tipping points "connectors," "mavens," and "salespersons" of the reform begin major efforts to review and revise state policies and create new criteria for local and state adoptions of instructional materials. With revision of standards, states also would initiate changes in assessments. Publishers would begin developing new editions of core and supplemental programs. Through this entire period, professional development of science teachers would continue.

Districts begin the process of selecting and implementing materials as they become available. Professional development aligned with the new programs is ongoing. The central goal of this phase is to revise local, state, and national policies; develop new school science programs; and align teaching practices with the goals of the reform.

By the end of this phase, states would have new standards and assessments; new teacher certification requirements would be in place; new instructional materials for core and supplemental programs would be available; and the professional development of teachers would be aligned with the new priorities and would be ongoing. This phase likely would present the most difficulty, as business leaders, policy makers, and educators will directly confront resistance to change and criticism of the new initiatives and changes in policies, programs, and practices.

Sustaining the Reform: Building Local Capacity for a National Purpose

The work of this phase would be concentrated in the final two years of the decade. In the next phase, work will concentrate on building local capacity for ongoing improvement of science and technology education at the district level. These efforts concentrate on a phase-out of dependence on external funds for the reform efforts and phase-in of school districts' use of resources in response to the new advances in science and technology and implied changes for the school programs.

Evaluating the Reform: Monitoring and Adjusting to Change

Evaluation will involve continuous feedback about the work and changes in content and curriculum, teachers and teaching, and assessments and accountability. Clearly, there will be feedback during all phases. The feedback will inform judgments about the modules and issues associated with their implementation and the professional development of teachers.

Evaluations and feedback will be conducted and available at the school district, state, national, and even international levels. School districts and states will implement their own evaluations. Results from NAEP, TIMSS, and PISA also will provide results from national and international levels.

Concluding Discussion

We have broad consensus on the goal of achieving scientific literacy for all learners, and the Standards, Benchmarks, and new common core standards provide policies that clarify the content and dimensions of our goal. It should be clear that there are options and opportunities to improve science programs and teaching practices. We must all assume responsibility for confronting the next challenges as we move toward our goal of achieving scientific literacy for all learners.

Fulfilling national aspirations has long been a function of science teaching, and curriculum materials have been a central component that helped science teachers attain national priorities. The Sputnik era serves as a prime example of this observation.

Among the accomplishments of the Sputnik era, we have organizations such as BSCS, which have the history and reputation of addressing the complex challenges of designing and developing innovative curriculum materials. It is time to set aside the old idea that science teachers and school districts can develop curriculum materials for entire courses with a few weeks of summer work. I see this change as fundamental as we enter a new era of curriculum reform.

Business and industry have signaled the need for curriculum reform in science education. Priorities associated with the No Child Left Behind legislation centered on basic literacy and mathematics. Some of these priorities are being addressed. Science and technology must now become a new priority because the contributions from science will provide the basis for higher levels of achievement in the knowledge, values, skills, and abilities required for the 21st century. The latter represents the national aspirations for this era. Achieving workforce competencies will require more than single initiatives that center on isolated components of the educational system. Rather, achieving workforce competencies will take coherent and coordinated efforts distributed across the key components of education, and we can begin with curriculum materials designed for science teachers.

The United States faces large, complex problems that require radical responses. Fifty years ago, the Sputnik challenge galvanized the nation in a way every citizen could understand. We need a similar sense of urgency and mission today. Both the challenges and our nation's response must be understood by every citizen. The purposes are clear: maintaining the United States' position as a global economic presence and addressing issues associated with climate change and energy resources. Now we must address the need for curriculum reform so that science and technology education once again fulfill national aspirations.

Having stated these recommendations, I will note some important features. First, my recommendations center on critical leverage points to address immediate and long-term problems. Second, the direct implication for federal policy is financial support versus unfunded mandates, requests for cooperation, general recommendations to state and local governments, or appeals for support from business and industry. Third, priorities include multiple and coordinated efforts among, for example, the U.S. Department of Education, the National Science Foundation, the National Institutes of Health, and other agencies. Fourth, the initiatives should build on current research, such as *How Students Learn: Science in the Classroom* (Donovan and Bransford 2005), *America's Lab Report* (NRC 2006), and *Taking Science to School* (NRC 2007). Finally, policy makers can support these priorities from a nonpartisan perspective. It is in the United States' interest to achieve higher levels of scientific literacy.

Epilogue

Science Teachers as 21st-Century Leaders

Since early in the 21st century, when I began preparing the lectures that form the heart of this book, there have been numerous and varied reports regarding the education system, with recommendations directed toward the ultimate aim of improving student achievement. Science teachers have heard about schools making "adequate yearly progress," systems closing the "achievement gap," and the country needing to "move to the top" on international assessments. Achievement on national assessments and state tests have come to dominate perceptions by the education community, including science teachers. Regardless of the phrases used to express the aims, the goals require attaining higher levels of achievement for all students.

The perspective in this book has centered on science teachers and issues closely related to science teaching and student learning. For example, I addressed themes such as fulfilling the goal of scientific literacy, reforming school science programs, teaching science as inquiry, and improving science teachers' knowledge and skills. All of these themes relate directly to the instructional core and, in the end, student achievement. Before turning to leadership, let me briefly restate and clarify the challenges from the first chapter.

21st-Century Challenges

There are five challenges that the science education community has to confront as we enter the second decade of the 21st century. Regardless of the current terms, I argue that these challenges and the responses are at the heart of science teaching, as they have been in the past and will continue to be in the future.

The challenges are

- achieving scientific literacy,
- reforming science programs,
- teaching science as inquiry,

- improving science teachers' knowledge and skills, and
- attaining higher levels of achievement for all students.

Achieving Scientific Literacy

The perspective I have placed on scientific literacy gives emphasis to a general education for all students as future citizens. The perspective differs from the traditional emphasis on science foundations. Incorporating the emphasis on scientific literacy as described in this book presents one of the most significant challenges to science education.

Reforming Science Programs

This perspective incorporates two things: first, the research on student learning and the implications for science teaching; second, recommendations that curriculum include current and future life situations and work contexts (for example, career, health, environment, resources, hazards, and research and development in science and technology).

Teaching Science as Inquiry

For too long this perspective has been viewed and presented as an either/or dichotomy for science teachers. Either you teach science for traditional science knowledge, or you teach science as inquiry. Again, I make three points. First, science teachers do not have to take an either/or view; they can take the both/and perspective. Second, teaching science as inquiry aligns with contemporary research on teaching and learning. Finally, teaching science as inquiry accommodates the contemporary goal of preparing students for life and work. That is, 21st-century workforce objectives can be attained by teaching science as inquiry.

Providing Professional Development

The changes included in the aforementioned discussion imply increasing science teachers' knowledge and skills. One should note that this sentence recommends increasing both knowledge and skills. Professional development often centers on acquiring knowledge without attention to the skills and strategies of science teachers' teaching.

Improving Student Achievement

The introduction addressed this challenge, as did my continued emphasis on the instructional core. This perspective is, after all, the change that science teachers strive for, districts want, and the nation needs.

A New Answer to the Sisyphean Question in Science Education

Hardly a day passes without a new report signaling the need for education reform. Throughout our history, society has continually demonstrated the need for education reform, so in one sense there is nothing new in the early decades of the 21st century. Contemporary calls for reform consistently reference reactions during the Sputnik era. The need for a response is accurate. However, it is now more than 50 years since Sputnik, and as the needs of society are different, so must the science education community's response be different.

All of us—educators, scientists, policy makers, and the business community—must ask and answer the perennial Sisyphean question for science education: What should citizens know, value, and be able to do in preparation for life and work in the 21st century?

Themes and discussions in this book have described my responses to the Sisyphean question. I have tried to address the issue of reform at the most fundamental level, that of the science teacher and classroom instruction. Now, I turn to the challenge of science teachers as 21st-century leaders.

Leadership

Leadership as discussed here is not that of a single great person. My point of view includes a majority of individuals in science education who can provide leadership, especially those with responsibilities for the teaching of science.

The vast and varied structure of science education and the scale and complicated nature of contemporary reform suggest that distributed leadership is essential. All individuals associated with science education must contribute to the common purpose of achieving scientific literacy for all students.

The themes of leadership and responsibility in reforming science education include the roles of teacher educators, science coordinators, science education researchers, and classroom teachers. Science teachers clearly have the greatest burden and heaviest responsibility for reform. Science teachers should not perceive this judgment as yet another form of teacher bashing. Rather, my comments are grounded in recognition of the essential and fundamental position of teachers in education, the need for reform, and a deep and sincere compassion for science teachers' difficult task.

Defining Leadership

Leadership is difficult to define. Several definitions are provided on the next page. Common themes unite under the theme of leadership. First, leadership involves relationships with other individuals. Second, leadership has the intention of achieving a goal common to the group. I can paraphrase an earlier publication in which I defined leadership as an individual ability to work with others to improve science teaching and student learning while achieving the goal of scientific literacy (Bybee 1993).

Definitions of Leadership

Leadership is making things happen or not happen.

Leadership is getting others to do what they ought to do, and like it.

Leadership is making people think things are possible that they didn't think were possible.

Leadership is getting people to be better than they think they are or can be.

Leadership is inspiring hope and confidence in others to accomplish purposes they think are impossible.

Leadership is perceiving what is needed and right and knowing how to mobilize people and resources to accomplish these goals.

Leadership is creating options and opportunities, clarifying problems and choices, building morale and coalitions, and providing a vision and possibilities of something better than currently exists.

Leadership is empowering and liberating people to become leaders in their own right.

Note that this definition includes a majority of individuals within the science education community but has particular relevance for science teachers.

The Responsibility and Leadership of Science Teaching

This section describes two related ideas. The first idea is that science teachers are ultimately responsible for what students learn in their classrooms, and the second is that leadership in the reform of science education arises from the recognition and acceptance of science teachers for their role in reform.

Science teachers have a tremendous responsibility. They must comprehend the complex concepts and processes of the sciences; they must incorporate into their program the processes that scientists use; and they must understand the intricate role that science plays in our society. And there is even more. Science teachers are also accountable for their students learning those same concepts, processes, and interactions of science and society. The responsibility is great. And, understandably, there are periods of doubt, vulnerability, loneliness, and anxiety about the significant trust society places in science teachers for the education of its youth.

Science teachers also are the leaders for science education in their classrooms, schools, and communities. This is the nature of teaching and the requirements of education. Science teachers themselves say, "When I close my door, I do as I know best in the classroom." This statement belies the dual ideas of responsibility and leadership. In the conduct of a lesson or a class period, science teachers are generally alone and solely responsible for what students learn.

Several facts ought to be made clear about the responsibilities of science teaching. First, the science teacher's responsibilities cannot be delegated. There is no place or person to turn to for the delegation of responsibility for students learning science. The classroom is the single channel through which students flow with their myriad needs, wants, disabilities, interests, misconceptions, naïve theories, and social agendas. Science teachers cannot wait for agreement on what must be done; they must make the many decisions that result in lessons being more or less effective, engage the learners, and maintain a positive climate within the classroom and school. Because most science teachers are alone in their classrooms, they can neither abdicate nor delegate responsibility. Other teachers, school administrators, and parents cannot assume the tasks of science teaching.

Second, the science teacher's responsibility is to all the students. Students come through the classroom door with a variety of challenges. From these students, science teachers must draw their strength and respond to the challenges with accurate and constructive strategies that enhance learning. If the science teacher constantly reassures a student with misconceptions that all is well, if the teacher answers all parents' concerns with an air of infallibility, or, worst of all, if the science teacher is not informed, he or she has failed the students. Science teachers cannot yield to the difficulties of diverse student groups; they must overcome, inform, and educate. Sometimes educating students requires risk, trying something new and unique with the hope that students will benefit.

Third, the science teacher's responsibility is to the community as well as the students. As though the duties of classroom teaching were not enough, science teachers must in many cases assume duties that require educating colleagues, administrators, school boards, and parents about science and science education. To do this, one must have an understanding of frontiers of science and trends and issues in science education, as well as some awareness of the political dynamics of the schools, communities, and nations.

Few in the contemporary reform of American education have realized the basic fact that what makes education work is inside the classroom: the sound development of science teachers as responsible leaders who can carry out their duties. The final purpose of the exercise of reforming science education is to provide science teachers with the means of doing the things that will foster the scientific literacy of students and the highest aspirations of society.

Empowering Science Teachers

Empowering science teachers is a prominent theme in the contemporary reform of science education (Spector 1989; Nyberg 1990). Teacher empowerment is an interesting contrast with the phrase popular in the 1960s: teacher-proof programs. A teacher-proof curriculum was a set of materials designed to enhance student learning independent of, or in spite of, the science teacher. As early as 1965, in

The Genius of American Education, Lawrence Cremin pointed out the misguided nature of a teacher-proof curriculum. As part of a discussion of education reform, Cremin suggested that reformers rightfully had a concern about contemporary teachers and teaching. But he saw their solution of designing materials impervious to misuse as flawed. Cremin gives advice that is appropriate for any generation of leaders and reformers, especially those teacher educators and science coordinators who are directly responsible for the professional development of science teachers.

> *But education is too significant and dynamic an enterprise to be left to mere technicians; and we might as well begin now the prodigious task of preparing men and women who understand not only the substance of what they are teaching but also the theories behind the particular strategies they employ to convey that substance. A society committed to the continuing intellectual, aesthetic, and moral growth of all its members can ill afford less on the part of those who undertake to teach.* (Cremin 1965, p. 57)

The quotation notes the disparity between the intentions of teacher-proof curriculum materials and the extent to which that goal was achieved. The teacher-proof approach to curriculum did not work.

In the wake of teacher-proof materials, teacher-dependent materials emerged. Although the origins may be in well-meaning responses to science teacher questions for instructional activities and materials that they can "use on Monday," the result has been a dependency that has become a countervailing force to Cremin's point (and also my themes in this book) of preparing men and women who understand both the substance of science teaching and the theories behind the particular strategies they employ. It is past time to reduce teachers' dependence on short-term quick fixes and develop long-term knowledge and skills that will indeed empower their science teaching.

Science teachers could, at this point, accurately ask—What are the qualities of empowerment? In a book titled *Leaders: The Strategies for Taking Charge*, Warren Bennis and Burt Nanus (1985) described several dimensions of empowerment. One of the first dimensions is *significance*. Effective leaders create a vision that makes others feel as though they make a difference. Of critical importance, the significance has substance and transcends the superficial significance of slogans. The individuals are, for example, translating the vision into innovative science programs and sustaining the new programs. Science teachers indeed would be making a difference in the education of their students.

A second dimension of empowerment involves developing new *knowledge*, *skills*, and *beliefs*. This results in greater competence and a sense of mastery. Third, empowerment provides a sense of *community*. For example, when all the science teachers in a school system have the common purpose of improving science education K–12, and they cooperate in achieving that purpose, they develop a sense of community and collegiality.

Empowering people results in their greater *enjoyment* of their work. This is a fourth dimension. Outdated theories of motivation and leadership suggested that only rewards and punishments of individuals could achieve desired results. Contemporary theories of motivation and leadership recognize that individuals have higher needs, and those include such motivations as needing to know and understand, engaging in meaningful work, and developing a personal and professional efficacy.

My theme of *responsibility* suggests a final dimension of empowerment. A critical aspect of empowerment is assuming the responsibilities for achieving the tasks. Science teachers must realize that empowerment and responsibility are two sides of the same coin. So, with empowerment, teachers have the responsibility to attain higher levels of student achievement. Science teachers also have to accept the consequence that with their relying on how-to activities, refusing to understand the larger education system, and resisting information because it is not relevant to their "real world," they relinquish power and avoid their responsibility for improving science education.

Leadership Requires a Vision and a Plan

One of the consistent requirements of education leadership is that leaders have both a vision and a plan. The point here is that teachers have *both* a vision and a plan. Many individuals have ideas for a better science education, but when asked about implementing their ideas they lack short-term tactics or long-term strategies. On the other hand, one no doubt has encountered individuals who have suggestions about how to manage things, but the individual has no goals or objectives, except better management. Leadership requires both a vision and strategies.

Vision

Leaders with vision may, for example, have a long-term perspective, see large systemic issues, present future scenarios, or discern fundamental problems and present possible solutions. As leaders, teachers do not spend time and energy assigning blame for problems. Depending on their situations, leaders have diverse ways of clarifying a vision. Some may do so in speeches, others in articles, and still others in policies. One leader's vision may unify a group, organization, or community; another's vision may set priorities or resolve conflicts among constituencies. A leader's vision likely will have many sources and result from extensive review and careful thought. This is especially true in today's complex education system.

Leadership in science education extends from science teachers to the secretary of education and the president. It does not reside with only a few people in key positions. Science education consists of numerous systems and subsystems, each with individuals who have power, constituents, and goals that contribute to a better science education for students. Not every member of the science education community can or should be involved in constructing international assessments, developing curriculum materials, presenting the arguments for scientific inquiry, defending the integrity of science, or providing professional development. But all of us do have our roles and responsibilities that relate to these and many other leadership opportunities, and that is what will ultimately make a difference for students.

The purpose of science education is for all students to achieve high levels of scientific literacy. Such a broad and, I would argue, deep perspective touches critical components of the science education system at the national, state, and local levels. The fundamental purpose is comprehensive and inclusive. This is the vision required of science education leaders in the 21st century.

Although visions of "science for *all* students," "*no* student left behind," and "race to the top" have become part of a state's review of policies, standards, and every school district's discussions of science programs, education leaders should point out the fact that such statements explicitly highlight equity. By "all students," we mean *all students*. The goal should be clear and unambiguous. Achieving the goal presents a complex array of problems. Now teacher leaders are required to restate and renew our efforts to make sure that all students have adequate and appropriate opportunities to learn science. I point to the issue of equity—and its frequently cited countervailing but paradoxical force, excellence—because it pervades all of education, not just science education.

Contemporary justification for a vision of improved science education resides in themes such as innovation and the economy, basic skills for the workforce, environmental quality, use of natural resources, and energy efficiency. Such themes differ from earlier justifications such as the space race and a nation at risk. In many respects, the economic rationale has emerged from the realization that the U.S. economy is part of a global economy and that the education level of our citizens influences the rate and direction of the country's economic progress.

Plans

The complement to a vision of science for all students is having a plan to enact that vision within the leader's context. The vision centers on students learning science. So how can we think about the contexts within which leaders should work, identify important initiatives, and formulate plans for constructive reform?

To enhance all students' scientific literacy, we must focus on the interactions at the instructional core—between teachers and students, especially those interactions that enhance learning. This, I believe, is a critically important education perspective. Enhancing learning includes placing curriculum materials, instructional strategies, classroom assessments, and continuous professional development in the foreground of the leader's vision. This perspective centers on the instructional core and clearly contrasts with contemporary political issues such as school choice, charter schools, and vouchers as means of higher levels of student achievement.

Leaders work in increasingly complex education systems. The time has passed when, for example, a leader could facilitate the selection of curriculum materials and trust that all would be well with their use and that, ultimately, there would be higher levels of student achievement. Now the complex system of science education includes political, economic, and social factors, as well as education issues. Effective leaders must recognize the multiple factors, varied components, and different aspects of the education system as they implement their plans.

Providing Leadership as a Science Teacher

Leadership qualities are sometimes attributed to individuals because of personality; we say they have charisma. In some cases, people are leaders because they have unique abilities that qualify them to lead. In other cases, there are people who are leaders because they are in positions of power and authority. It would be nice if, as a summary, I could say that science teachers were leaders because they had all of these qualities: charisma, competence, and control. The truth is, few teachers possess all of these qualities. Teachers do have some power and control because of their position, and one assumes they are competent in their scientific knowledge, teaching methods, planning skills, classroom organization, and management. Individuals may or may not have the enthusiasm and personality for charismatic leadership.

Leadership in the Classroom

Leadership in the science classroom involves developing a climate that sustains efficient and effective work by the classroom group while fulfilling personal needs and education goals. In other words, leadership requires management of the total classroom—science content, the physical environment, individual student needs, and the students as a group. However, it would be misleading to leave the impression that leadership is solely a function of the teacher. One needs only to recall the pleasant surprises that await substitute teachers to realize that leadership can originate from the ranks of students. In less extreme examples, leadership is commonly seen when students work in small groups on science activities. The two major functions of classroom leadership are facilitation and maintenance.

The tasks of facilitation that contribute to effective leadership include

- establishing policies and standards for the classroom,
- coordinating work procedures among students and groups,
- improving the classroom climate through cooperative problem solving, and
- modifying physical conditions in the classroom.

As leaders, science teachers neither coerce nor persuade; rather, they facilitate classroom unity by setting policies, procedures, and conditions through cooperative interaction with the student group. This cooperation has a tremendous positive influence on individual behavior. When the classroom group has the ability to resolve problems and make decisions, many of the everyday problems of organization and management are avoided.

Even though the classroom group works together and continues to fulfill education goals, there are inevitably times when management problems will arise. Schedule changes, all-school activities, and unplanned custodial work can all cause changes in the physical environment, classroom climate, or group composition. Such situations often require fast action by the teacher to maintain the classroom group. Some maintenance functions of leadership include

- sustaining morale,
- resolving conflicts,
- restructuring groups changed due to outside factors, and
- reducing students' anxiety and fear.

Maintaining the classroom group in the face of changing and sometimes adverse conditions requires flexibility and adaptability by the science teacher.

A second and essential dimension of leadership is in the school and community. There are different qualities of leadership at this level. I return to the idea of a vision, in this case a vision of what science education can and should be. The leader of science education works to effect change and improvement. Effective leadership develops hope, builds confidence, and generates new attitudes about the possibilities of science education. Leaders bring people together and develop plans that will overcome the costs, constraints, and risks that attend the reform of school science programs. Leadership by science teachers involves getting colleagues, administrators, and school boards to do something they are reluctant to do—change.

Science teachers who lead don't expect leadership to emerge from others such as administrators. There is a long history of change that originates not in official policies, but by individuals who have dreams of a better world. Only later are the dreams written into policies, laws, and programs. The civil rights movement, for example, originated with individuals such as Rosa Parks, was taken up by Martin Luther King, and only later became the law of the land.

Providing leadership in science education can begin by assuming the responsibility of improving your own science programs. Most science teachers do this on a continuing basis. But what is your vision? Where are you taking your students? What should your students be learning about science? I must say that even though you have a responsibility to provide leadership, it is not easy. I will also say that in the end, it is worth the effort.

Confronting the Paradoxes of Leadership

Science teachers assuming the responsibilities of leadership will inevitably confront persons, situation, and actions that are apparently contradictory. These are, by definition, paradoxes. A paradox is a statement or situation that on the surface seems contradictory. Earlier I mentioned an often-heard paradox in education—equity for all students versus excellence for a few students. Paradoxes can be resolved. For example, a leader must maintain continuity with past science programs while initiating changes with new curricula. Leaders often express paradoxes as tensions, contradictory directions, or conflicting issues. However, the forces seen as countervailing elements of a paradox may not be as diametrically opposed as they seem; in fact, the apparently contradictory goals may reinforce each other. Leaders must master the paradoxes they confront. Let me describe several paradoxes faced by education leaders.

Science Education's Paradoxes

One of the classic paradoxes that science teachers may confront is encouraging change in science programs and practices while simultaneously supporting maintenance of past programs and practices. The resolution may center on maintaining stability in the major concepts of life science while adopting a new inquiry-oriented biology program or changing the sequence of biology, chemistry, and physics courses to some we have termed *physics first*—that is, a program of physics, chemistry, and biology, in that order.

A second example of a paradox that leaders face involves having a clear direction while being open and flexible to suggestions for different directions. The resolution here may center on ultimate and proximate goals. In the long-term, the leader may have a consistent view of the goal he or she wants to attain. However, in the short-term the leader may have to accept changes that only partially represent the final goal. In the interim the leader remains flexible and open to new ways of achieving the vision.

In prior discussions for the five themes, I suggested some issues the leaders may confront as paradoxes. Achieving higher levels of scientific literacy may be presented as a choice between teaching science content or teaching social contexts such as climate change or energy efficiency. Changing school science programs may be presented as the difficulty of bridging theory and practice. Teaching science as inquiry has consistently met resistance from those who see the choice as either content or process. Providing professional development has emphasized science content as opposed to pedagogy, and student achievement has been reduced to individual test results versus national achievement and international comparisons. All of these paradoxes can be resolved.

Recognizing the Politics

Along with the central importance of resolving the tensions of paradoxes, I would list the importance of a leader's ability to recognize and address the political realities of education work. The leader has to recognize that initiating changes means addressing the politics. All issues of improving science achievement are not solely related to education. Indeed, it may be the case that all education issues ultimately are political issues. The paradox embedded here can be stated as achieving education goals while addressing political realities. I have found that either/or thinking often expresses the paradox, while both/and thinking provides insights into the resolutions.

Experience teaches another lesson for those in leadership positions. If you are leading, you cannot avoid conflict and controversy. Also, the larger the system and greater the change, the more controversy you will experience. It can be thought of as the paradox "achieving your goals requires enduring criticism." And the criticism often is unfair and personal.

Concluding Discussion

Our discussion here, while this 21st century is still young, presents the occasion to review trends and issues that science teachers will encounter as leaders. What is common to the work of leaders? I proposed establishing a clear and consistent vision combined with a practical and workable plan. The vision and plan will get the leader started in directions that may involve curriculum reform, instructional improvement, or alignment of assessments. One crucial point that I made is that leaders must hone their ability to realize and resolve paradoxes as they execute their plans. The paradoxes have been referred to as tensions, critical problems, and even absurdities. Regardless, effective leadership requires the resolution of paradoxes such as initiating bold new programs while maintaining established past traditions, or fulfilling a national mandate such as NCLB or common core standards while incorporating a local agenda. One of the most disheartening paradoxes is the reality of achieving the established vision and enduring criticism rather than receiving a reward for attaining the goal. Given this view of leadership in science education, I described several themes that leaders will confront in the first decades of the 21st century.

I identified five themes that will directly or indirectly influence science education leaders. The themes are achieving scientific literacy, reforming science programs, teaching science as inquiry, improving science teachers' knowledge and skills, and attaining higher levels of achievement for all students.

I made the case that the instructional core—curriculum instruction, assessment, and professional development—is where our time, money, and effort should be focused. Improvements in the instructional core will, in both the short- and long-terms, bring the greatest advances toward scientific literacy for all students. Among the crucial aspects of the instructional core, one has to include the understanding of scientific inquiry by classroom teachers and their subsequent efforts to help students develop the cognitive abilities and conceptual understandings aligned with this aspect of science education.

I see the continued need for professional development. Relative to this theme, I note that it should be integrated with other meaningful activities, such as curriculum reform.

Within each of the five themes, one can easily identify other challenges that educational leaders must confront. Such is the reality of reform in science education. Each leader has his or her individual and unique circumstances, whether in the science classroom, superintendent's office, board of education or public agencies, and private organizations at the state and national levels. Regardless of the unique situations, we are all part of a larger science education system that strives toward a goal of attaining higher levels of scientific literacy for all of our students.

Appendix

APPENDIX I: ACID RAIN

PISA 2006—Science—Released Unit and Items

Below is a photo of statues called Caryatids that were built on the Acropolis in Athens more than 2500 years ago. The statues are made of a type of rock called marble. Marble is composed of calcium carbonate.

In 1980, the original statues were transferred inside the museum of the Acropolis and were replaced by replicas. The original statues were being eaten away by acid rain.

Photo by Maxhomand for iStockphoto.

Question 2: ACID RAIN
Difficulty: 506

Normal rain is slightly acidic because it has absorbed some carbon dioxide from the air. Acid rain is more acidic than normal rain because it has absorbed gases like sulfur oxides and nitrogen oxides as well.

Where do these sulfur oxides and nitrogen oxides in the air come from?

_____

_____

_____

The effect of acid rain on marble can be modelled by placing chips of marble in vinegar overnight. Vinegar and acid rain have about the same acidity level. When a marble chip is placed in vinegar, bubbles of gas form. The mass of the dry marble chip can be found before and after the experiment.

Question 3: ACID RAIN
Difficulty: 460

A marble chip has a mass of 2.0 grams before being immersed in vinegar overnight. The chip is removed and dried the next day. What will the mass of the dried marble chip be?

   A. Less than 2.0 grams
   B. Exactly 2.0 grams
   C. Between 2.0 and 2.4 grams
   D. More than 2.4 grams

Question 5: ACID RAIN
Difficulty: Partial Credit—513, Full Credit—717

Students who did this experiment also placed marble chips in pure (distilled) water overnight.

Explain why the students included this step in their experiment.

_____

_____

Question 10N: ACID RAIN

How much interest do you have in the following information?

Tick only one box in each row.

	High Interest	Medium Interest	Low Interest	No Interest
d) Knowing which human activities contribute most to acid rain	☐	☐	☐	☐
e) Learning about technologies that minimise the emission of gases that cause acid rain	☐	☐	☐	☐
f) Understanding the methods used to repair buildings damaged by acid rain	☐	☐	☐	☐

Question 10S: ACID RAIN

How much do you agree with the following statements?

Tick only one box in each row.

	Strongly Agree	Agree	Disagree	Strongly Agree
G) Preservation of ancient ruins should be based on scientific evidence concerning the causes of damage.	☐	☐	☐	☐
H) Statements about the causes of acid rain should be based on scientific research	☐	☐	☐	☐

Appendix II: GREENHOUSE

PISA 2006—Science—Released Unit and Items

Read the texts and answer the questions that follow.

The Greenhouse Effect: Fact or Fiction?

Living things need energy to survive. The energy that sustains life on the Earth comes from the Sun, which radiates energy into space because it is so hot. A tiny proportion of this energy reaches the Earth.

The Earth's atmosphere acts like a protective blanket over the surface of our planet, preventing the variations in temperature that would exist in an airless world.

Most of the radiated energy coming from the Sun passes through the Earth's atmosphere. The Earth absorbs some of this energy, and some is reflected back from the Earth's surface. Part of this reflected energy is absorbed by the atmosphere.

As a result of this the average temperature above the Earth's surface is higher than it would be if there were no atmosphere. The Earth's atmosphere has the same effect as a greenhouse, hence the term *greenhouse effect*.

The greenhouse effect is said to have become more pronounced during the twentieth century.

It is a fact that the average temperature of the Earth's atmosphere has increased. In newspapers and periodicals the increased carbon dioxide emission is often stated as the main source of the temperature rise in the twentieth century.

A student named André becomes interested in the possible relationship between the average temperature of the Earth's atmosphere and the carbon dioxide emission on the Earth.

In a library he comes across the following two graphs.

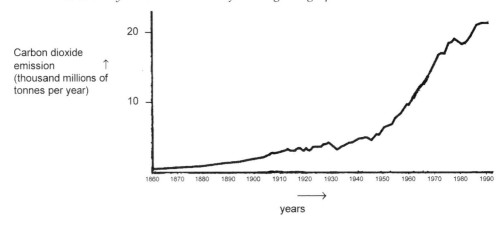

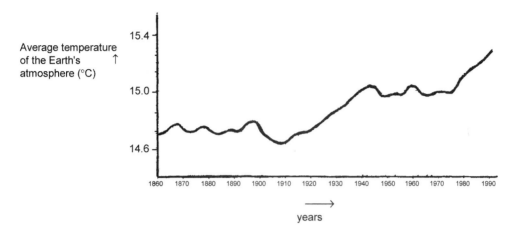

André concludes from these two graphs that it is certain that the increase in the average temperature of the Earth's atmosphere is due to the increase in the carbon dioxide emission.

Question 3: GREENHOUSE

Difficulty: 529

What is it about the graphs that supports André's conclusion?

_____

_____

_____

Question 4: GREENHOUSE

Difficulty: Partial Credit—568, Full Credit—659

Another student, Jeanne, disagrees with André's conclusion. She compares the two graphs and says that some parts of the graphs do not support his conclusion.

Give an example of a part of the graphs that does not support André's conclusion. Explain your answer.

_____

_____

_____

Question 5: GREENHOUSE

Difficulty: 709

André persists in his conclusion that the average temperature rise of the Earth's atmosphere is caused by the increase in the carbon dioxide emission. But Jeanne thinks that his conclusion is premature. She says: "Before accepting this conclusion, you must be sure that other factors that could influence the greenhouse effect are constant."

Name one of the factors that Jeanne means.

_____

_____

_____

Source: Organisation for Economic Cooperation and Development (OECD). 2009. *PISA 2009 assessment framework: Key competencies in reading, mathematics, and science.* Paris: OECD.

References

Abell, S. K., and N. Lederman, eds. 2007. *Handbook of research on science education.* New Jersey: Lawrence Erlbaum Associates.

Adelson, R. 2004. Instruction versus exploration in science learning: Recent psychological research calls "discovery learning" into question. *Monitor on Psychology* 35 (6): 34.

American Association for the Advancement of Science (AAAS). 1989. *Science for all Americans: A Project 2061 report on goals in science, mathematics, and technology.* Washington, DC: AAAS.

American Association for the Advancement of Science (AAAS). 1993. *Benchmarks for science literacy.* New York: Oxford University Press.

American Association for the Advancement of Science (AAAS). 1995. Common ground: Benchmarks and national standards. *2061 Today* 5 (1): 1–2.

American Association for the Advancement of Science (AAAS). 1999. *Middle grades science textbooks: A benchmarks-based evaluation.* Washington, DC: AAAS.

American Association for the Advancement of Science (AAAS). 2000. *High school biology textbooks: A benchmarks-based evaluation.* Washington, DC: AAAS.

American Association for the Advancement of Science (AAAS). 2001. *Resources for science literacy: Curriculum materials evaluation.* New York: Oxford University Press.

Atkin, J. M., P. Black, and J. Coffey, eds. 2001. *Classroom assessment and the national science education standards.* Washington, DC: National Academies Press.

Atkin, J. M., and R. Karplus. 1962. Discovery of invention? *Science Teacher* 29 (5): 45.

Autor, D., F. Levy, and R. J. Murnane. 2003. The skill content of recent technical change. *Quarterly Journal of Economics* 118 (4): 1279–1333.

Ball, D. L. 1996. Teaching learning and the mathematics reforms: What we think we know and what we need to learn. *Phi Delta Kappan* 77 (7): 500–506.

Bardeen, M., and L. Lederman. 1998. Coherence in science education. *Science* 281 (10 July): 178–179.

Barnett, C. 1998. Mathematics teaching cases as a catalyst for informed strategic inquiry. *Teaching and Teacher Education* 14 (1): 81–93.

Bass, H. 1998. The math education debates. Remarks developed from presentations to the Center for Science, Mathematics, and Engineering Education, Irvine, CA, and later expanded to the Research Symposium—Reflecting on the math wars: Perspectives on the role of research and researchers in the public discourse about mathematics education reform, NCTM meeting, Washington, DC.

Begley, S. *Wall Street Journal*. 2004a. The Best Ways to Make School Children Learn? We Just Don't Know. December 10.

Begley, S. *Wall Street Journal*. 2004b. To Improve Education, We Need Clinical Trials to Show What Works. December 17.

Bennis, W., and B. Nanus. 1985. *Leaders: The strategies for taking change*. New York: Harper and Row.

Biological Sciences Curriculum Study (BSCS). 1960. BSCS committee plans program of original investigations for science-prone students. *BSCS Newsletter* 3 (May): 3.

Biological Sciences Curriculum Study (BSCS). 1961. Biological Investigations for secondary school students. *BSCS Newsletter* 7 (April): 4.

Biological Sciences Curriculum Study (BSCS). 1989. *New designs for elementary school science and health: A cooperative project of Biological Sciences Curriculum Study (BSCS) and International Business Machines (IBM)*. Dubuque, IA: Kendall/Hunt.

Biological Sciences Curriculum Study (BSCS). 1992. *Science for life and living*. Dubuque, IA: Kendall/Hunt.

Biological Sciences Curriculum Study (BSCS). 1993. *Developing biological literacy: A guide to developing secondary and post-secondary biology curricula*. Dubuque, IA: Kendall/Hunt.

Biological Sciences Curriculum Study (BSCS). 1999. *BSCS Science T.R.A.C.S.* Dubuque, IA: Kendall/Hunt.

Biological Sciences Curriculum Study (BSCS). 2000. *Making sense of integrated science: A guide for high schools*. Colorado Springs, CO: BSCS.

Biological Sciences Curriculum Study (BSCS), the SCI Center. 2002. *Profiles in science: A guide to NSF-funded high school instructional materials*. Colorado Springs, CO: BSCS.

Biological Sciences Curriculum Study (BSCS). 2005. *BSCS science: An inquiry approach*. Dubuque, IA: Kendall/Hunt.

Biological Sciences Curriculum Study (BSCS). 2007. *A decade of action: Sustaining global competitiveness*. Executive Summary. Colorado Springs, CO: BSCS.

Blank, R. K., and D. Langesen. 2001. *State indicators of science and mathematics education: State-by-state trends and new indicators from the 1999–2000 school year*. Washington, DC: Council of Chief State School Officers.

Boekel, N. J., and R. W. Bybee. 1973. The influence of science in a program for educable mentally retarded. In *Becoming a better elementary science teacher,* ed. Robert B. Sund, 123–136. Columbus, OH: Charles E. Merrill, A. Bell & Howell.

Brandwein, P. F. 1962. *Elements in a strategy for teaching science in elementary school: The Burton lecture.* New York: Harcourt, Brace, & World.

Brandwein, P. 1965. *Substance, structure, and style in the teaching of science.* New York: Harcourt Brace Jovanovich.

Brandwein, P., J. Metzner, E. Morholt, A. Roe, and W. Rosen. 1962. Teaching high school biology: A guide to working with potential biologists. *Biological Sciences Curriculum Study Bulletin No. 2.* Washington, DC: American Institute of Biological Sciences, Jovanovich.

Bransford, J. D., A. L. Brown, and R. R. Cocking, eds. 1999. *How people learn: Brain, mind, experience, and school.* Washington, DC: National Academies Press.

Bransford, J. D., A. L. Brown, and R. R. Cocking, eds. 2000. *How people learn: Brain, mind, experience, and school.* Exp. ed. Washington, DC: National Academies Press.

Brown, L., et al. 1990. *State of the world: 1990.* New York: W. W. Norton.

Bruner, J. 1960. *The process of education.* New York: Vintage Books.

Bybee, R. W. 1972. Science in a silent world. *Science Activities* 6: 25–54.

Bybee, R. W. 1979a. Science education and the emerging ecological society. *Science Education* 63 (1): 95–109.

Bybee, R. W. 1979b. Science education for an ecological society. *American Biology Teacher* 41 (3): 154–163.

Bybee, R. W. 1979c. Science education policies for an ecological society: Aims and goals. *Science Education* 63 (2): 245–255.

Bybee, R. W. 1984. *Human ecology: A perspective for biology education.* Monograph series II. Reston, VA: National Association of Biology Teachers (NABT).

Bybee, R. W. 1985. The Sisyphean question in science education: What should the scientifically and technologically literate person know, value, and do—as a citizen? In *Science technology society: 1985 yearbook of the National Science Teachers Association,* ed. R. Bybee, 79–93. Arlington, VA: National Science Teachers Association.

Bybee, R. W. 1991. Planet Earth in crisis: How should science educators respond? *American Biology Teacher* 53 (3): 146–153.

Bybee, R. W. 1993. *Reforming science education: Social perspectives and personal reflections.* New York: Teachers College Press.

Bybee, R. W. 1997. *Achieving scientific literacy: From purposes to practices.* Portsmouth, NH: Heinmann.

Bybee, R. W. 2001. Unintentional consequences of an acceptable evaluation. *The Biology Teacher* 63 (1): 2.

Bybee, R. W., ed. 2002. *Learning science and the science of learning: Science educators' essay collection.* Arlington, VA: NSTA Press.

Bybee, R. W. 2003. Introducing urban ecosystem education into educational reform. In *Understanding urban ecosystems*, ed. A. Berkowitz, C. Nilon, and K. Hollweg, pp. 430–449. New York: Springer.

Bybee, R. W. 2005. Scientific inquiry and science teaching. *In Scientific inquiry and the nature of science: Implications for teaching, learning, and teacher education*, ed. L. B. Flick and N. G. Lederman, 1–14. Boston: Kluwer Academic Publishers.

Bybee, R. W. 2007. Do we need another sputnik? *The American Biology Teacher* 69 (8): 454–457.

Bybee, R. W. 2008. Scientific literacy, environmental issues, and PISA 2006: The 2008 Paul F-Brandwein lecture. *Journal of Science Education and Technology* 17: 566–585.

Bybee, R. W. 2009. The BSCS 5E Instructional Model and 21st century skills. Paper prepared for the Workshop on Exploring the Interaction of Science Education and the Development of 21st Century Skills, National Research Council, Washington, DC. *http://www7.nationalacademies.org/bose/21centskilluploads.html*.

Bybee, R. W., and G. DeBoer. 1993. Goals for the science curriculum. *Handbook of research in science teaching and learning*. New York: MacMillan.

Bybee, R. W., and B. McCrae, eds. 2009. *PISA Science 2006: Implications for science teachers and teaching*. Arlington, VA: NSTA Press.

Bybee, R. W., B. McCrae, and R. Laurie. 2009. PISA 2006: An assessment of scientific literacy. *Journal of Research in Science Teaching* 46 (8): 865–883.

Bybee, R. W., and R. B. Sund. 1982. *Piaget for educators*. Columbus, OH: Merrill.

Bybee, R. W., J. A. Taylor, A. L. Gardner, P. Van Scotter, P. J. Carlson, A. Westbrook, and N. Landes. 2006. *The BSCS 5E Instructional Model: Origins and effectiveness*. Washington, DC: National Institutes of Health Office of Science Education.

Cavanagh, S. *Education Week*. 2004. NCLB could alter science teaching. November 9.

Champagne, A. 1987. The psychological basis for a model of science instruction. Commissioned paper for IBM-supported design project. Colorado Springs, CO: BSCS.

Chen, Z., and D. Klahr. 1999. All other things being equal: Acquisition and transfer of the control of variables strategy. *Child Development* 70 (5): 1098–1120.

Cohen, D. K., and H. Hill. 1998. *Instructional policy and classroom performance: The mathematics reform in California*. CPRE Research Report Series RR-39. Philadelphia: Consortium for Policy Research in Education.

Conant, J. B. 1951. *Science and common sense: A world-famous scientist and educator explains how science works*. New Haven, CT: Yale University Press.

Conant, J. B., ed. 1957. *Harvard case histories in experimental science,* vol. 2. Cambridge, MA: Harvard University Press.

Costenson, K., and A. Lawson. 1986. Why isn't inquiry used in more classrooms? *American Biology Teacher* 48 (3): 150–158.

Cremin, L. 1965. *The genius of American education*. New York: Vintage Books.

DeBoer, G. E. 1991. *A history of ideas in science education*. New York: Teachers College Press.

DeBoer, G. 2000. Scientific literacy: Another look at its historical and contemporary meanings and its relationship to science education reform. *Journal of Research in Science Teaching* 37 (6): 582–601.

Dewey, J. 1910. Science as subject matter and as method. *Science* 31: 121–127.

Dewey, J. 1938. *Logic: The theory of inquiry*. New York: Henry Holt.

Dewey, J. 2005. *How we think*. 1910. Reprint, New York: Barnes and Noble.

Donovan, M., and J. Bransford, eds. 2005. *How students learn: Science in the classroom*. Washington, DC: National Academies Press.

Donovan, M. S., J. D. Bransford, and J. W. Pellegrino, eds. 1999. *How people learn: Bridging research and practice*. Washington, DC: National Academies Press.

Dow, P. 1991. *Schoolhouse politics*. Cambridge, MA: Harvard University Press.

Duschl, R., H. Schweingruber, and A. Shouse, eds. 2007. *Taking science to school: Learning and teaching science in grades K–8*. Washington, DC: National Academies Press.

Education Development Center (EDC). 2007. DC synthesis of research on the impact of inquiry science instruction (summary). *http://cse.edc.org/products/inquirysynth*.

Eisner, E. W. 1971. *Confronting curriculum reform*. Boston: Little, Brown.

Elmore, R. E. 2004. *School reform from the inside out: Policy, practice, and performance*. Cambridge, MA: Harvard Education Press.

Elmore, R. 2009. Improving the instructional core. In *Instructional rounds in education: A network approach to improving teaching and learning*, ed. E. U. City, R. Elmore, S. Fiarman, and L. Teite, 21–38. Cambridge, MA: Harvard Education Press.

Fuller, R. G. 2002. *A love of discovery: Science education—the second career of Robert Karplus*. New York: Kluwer Academic/Plenum Publishers.

Funk, C. 2002. External evaluation and follow-up report for the BSCS Keys to Science Institute 2001. BSCS ER 2002-C1 August.

Gladwell, M. 2002. *The tipping point*. Boston, MA: Little, Brown.

Glass, B. 1976. Reflections on the early days of BSCS. *BSCS Newsletter* 64: 3–4

Gonzales, P., J. C. Guzman, L. Partelow, E. Pahlke, D.C. Miller, L. Jocelyn et al. 2004. *Pursuing excellence: Fourth-grade mathematics and science achievement in the United States and other countries from the Trends in International Mathematics and Science Study (TIMMS) 2003 NCES 2005–007*. Washington, DC: National Center for Education Statistics.

Gonzales, P., E. Pahlke, J. C. Guzman, L. Partelow, D. Kastberg, L. Jocelyn et al. 2004. *Pursuing excellence: Eighth-grade mathematics and science achievement in the United States and other countries from the Trends in International Mathematics and Science Study (TIMMS) 2003 NCES 2005–007*. Washington, DC: National Center for Education Statistics.

Grobman, A. 1963. Quoted in American Association for the Advancement of Science. *The new school science: A report to school administrators on regional orientation conferences in science.* Publication No. 63-6, p. 27. Washington, DC.

Grobman, A. 1969. *The changing classroom: The role of the biological sciences curriculum study.* New York: Doubleday.

Hardin, G. 1968. The tragedy of the commons. *Science* 162 (December 13): 1243–1248.

Harms, N., and S. Kohl. 1980. Project synthesis. Final report submitted to the National Science Foundation. Boulder, CO: University of Colorado.

Harms, N., and R. Yager, eds. 1981. *What research says to the science teacher.* Vol. 3. Arlington, VA: National Science Teachers Association.

Harvard Committee. 1945. *General education in a free society.* Cambridge, MA: Harvard University Press.

Harvard University. 1889. Descriptive list of elementary physical experiments intended for use in preparing students for Harvard College. Cambridge, MA: Harvard University.

Hawkins, D. 1965. Messing about in science. *Science and Children* 2 (6): 5–9.

Holton, G. 1998. 1948: The new imperative for science literacy. *Journal of College Science Teaching* 28 (3): 181–185.

Houston, J. 2007. Future skill demands: From a corporate consultant perspective. Presentation at the Workshop on Research Evidence Related to Future Skill Demands, National Academies of Science, Washington, DC. *http://www7. nationalacademies.org/cfe/future_skill_demands_presentations.html.*

Hurd, P. D. 1958. Science literacy: Its meaning for American schools. *Educational Leadership* 16: 13–19.

Hurd, P., R. W. Bybee, J. B. Kahle, and R.Yager. 1980. Biology education in secondary schools of the United States. *The American Biology Teacher* 42 (7): 388–410.

Johnson, D., R. Johnson, and E. Holubec. 1986. *Circles of learning: Cooperation in the classroom.* Alexandria, VA: Association for Supervision and Curriculum Development.

Karplus, R. 1971. Three guidelines for elementary school science. *SCIS Newsletter* 20: 368–371. Berkeley, CA: Regents of the University of California.

Karplus, R. 1977. Science teaching and the development of reasoning. *Journal of Research in Science Teaching* 14 (2): 169–175.

Karplus, R., and H. D. Thier. 1967. *A new look at elementary school science.* Chicago: Rand McNally.

Klahr, D., Z. Chen, and E. E. Toth. 2001. Cognitive development and science education: Ships that pass in the night or beacons of mutual illumination? In *Cognition and instruction: 25 years of progress*, ed. S. M. Carver and D. Klahr, 75–119. Mahwah, NJ: Erlbaum.

Klahr, D., and J. Li. 2005. Cognitive research and elementary science instruction: From the laboratory, to the classroom, and back. *Journal of Science Education and Technology* 14 (2): 217–238.

Klahr, D., and M. Nigam. 2004. The equivalence of learning paths in early science instruction: Effects of direct instruction and discovery learning. *Psychological Science* 15 (10): 661–667.

Kliebard, H. 1994. Curriculum ferment in the 1890s. In *The future of education: Perspectives on national standards in America*, ed. M. Cobb, 17–39. New York: College Entrance Examination Board.

Koballa, T., A. Kemp, and R. Evans. 1997. The spectrum of scientific literacy. *The Science Teacher* 64 (7): 27–31.

Kollmuss, A., and J. Agyeman. 2002. Mind the gap: Why do people act environmentally and what are the barriers to pro-environmental behavior? *Environmental Education Research* 8 (3): 239–260.

Kuhn, T. S. 1970. *The structure of scientific revolutions.* Chicago: University of Chicago Press.

Kulm, G., and S. Malcom, eds. 1991. *Science assessment in the service of reform.* Washington, DC: American Association for the Advancement of Science.

Kyle, W. C., Jr., R. J. Bonnstetter, J. McCloskey, and B. A. Fults. 1985. What research says: Science through discovery: Students love it. *Science and Children* 23 (2): 39–41.

Lamb, T. A. 2001. Evaluation report for the 2001 Keys to Science Institute. BSCS ER 2001-05 August. Colorado Springs, CO.

Lamb, T. A. 2002a. Evaluation report for the 2002 Eisenhower National Clearinghouse Demonstration Site Coordinators Workshop. BSCS ER 2002-07 August. Colorado Springs, CO.

Lamb, T. A. 2002b. Evaluation report for the 2002 Keys to Science Institute. BSCS ER 2002-06 August. Colorado Springs, CO.

Lawson, A. E. 1995. *Science teaching and the development of thinking.* Belmont, CA: Wadsworth Publishing.

Lawson, A. E., M. Abraham, and J. Renner. 1989. *A theory of instruction: Using the learning cycle to teach science concepts and thinking skills: NARST Monograph Number One.* Manhattan, KS: National Association for Research in Science Teaching.

Lederman, N. G. 1992. Students' and teachers' conception of the nature of science: A review of the research. *Journal of Research in Science Teaching* 29: 331–359.

Lemke, M., A. Sen, E. Pahlke, L. Partelow, D. Miller, T. Williams, et al. 2004. *International outcomes of learning in mathematics literacy and problem solving: PISA 2003 results from the U.S. perspective (NCES 2005–003).* Washington, DC: National Center for Education Statistics, Department of Education.

Lemke, M., A. Sen, E. Pahlke, L. Partelow, D. Miller, T. Williams, D. Kastberg, and L. Jocelyn. 2005. *Instructional outcomes of learning mathematics literacy and problem solving.* Washington, DC: National Center for Education Statistics.

Levy, F., and R. J. Murnane. 2004. *The new division of labor: How computers are creating the next job market*. Princeton, NJ: Princeton University Press.

Linder, C., L. Ostman, and P. Wickman, eds. 2007. Promoting scientific literacy: Science education research in transaction. Proceedings of the Linnaeus Tercentenary Symposium, Uppsala University, Uppsala, Sweden.

Loucks-Horsley, S., P. W. Hewson, N. Love, and K. E. Stiles. 2003. *Designing professional development for teachers of science and mathematics*. 2nd ed. Thousand Oaks, CA: Corwin Press.

Marek, E. A., and A. M. L. Cavallo. 1997. *The learning cycle*. Portsmouth, NH: Heinemann.

Metzenberg, S. 1998. Testimony before the United States House of Representatives Committee on Science, Subcommittee on Basic Research, July 23.

Michaels, S., A. W. Shouse, and H. A. Schweingruber. 2008. *Ready, set, science: Putting research to work in K–8 Science Classrooms*. Washington, DC: National Academy of Sciences.

Millar, R. 2006. Twenty first century science: Insights from the design and implementation of a scientific literacy approach in school science. *International Journal of Science Education* 28 (13): 1499–1521.

Millar, R., and R. Driver. 1987. Beyond process. *Studies in Science Education* 14: 33–62.

Minner, D., A. J. Levy, and J. Century. 2010. Inquiry-based science instruction—what is it and does it matter? Results from a research synthesis years 1984–2002. *Journal of Research in Science Teaching* 47 (4): 474–496.

Morrone, M., K. Manci, and K. Carr. 2001. Development of a metric to test group differences in ecological knowledge as one component of environmental literacy. *Journal of Environmental Education* 32 (4): 33–42.

Muller, H. J. 1957. Man's place in living nature. *Scientific Monthly* 84: 252–254.

Mullis, I. V., M. O. Martin, T. A. Smith, K. D. Garden, K. Gregory, E. G. Gonzalez, S. J. Chrostowski, and K. M. O'Connor. 2001. *Assessment frameworks and specifications 2003: Science framework*. Instructional Association for the Evaluation of Education (IEA). Chestnut Hill, MA: Boston College.

Murnane, R. J., and E. Levy. 1996. *Teaching the new basic skills: Principles for educating children to thrive in a changing economy*. New York: Free Press.

National Academies. 2005. *Rising above the gathering storm: Energizing and employing America for a brighter economic future*. Washington, DC: National Academies Press.

National Assessment of Educational Progress (NAEP). 2007. *The nation's report card: 12th grade reading and mathematics 2005*. W. Grigg, P. Donahue, and G. Dion, U.S. Department of Education, National Center for Education Statistics. Washington, DC: U.S. Government Printing Office. *http://nces.ed.gov/nationsreportcard/pdf/main2005/20077468.pdf.*

National Assessment Governing Board (NAGB). 2005. *Science assessment and item specifications for the 2009 national assessment of educational progress*. Washington, DC: NAGB.

National Assessment Governing Board (NAGB). 2009. *Science framework for the 2009 national assessment of educational progress*. Washington, DC: NAGB.

National Education Association (NEA). 1894. *Report of the Committee of Ten on secondary school studies*. New York: American Book Company.

National Research Council (NRC). 1996. *National Science Education Standards*. Washington, DC: National Academies Press.

National Research Council (NRC). 2000. *Inquiry and the National Science Education Standards: A guide for teaching and learning*. Washington, DC: National Academies Press.

National Research Council (NRC). 2006. *America's lab report: Investigations in high school science*. Washington, DC: National Academies Press.

National Research Council (NRC). 2007. *Taking science to school: Learning and teaching science in grades K–8*. Washington, DC: National Academies Press.

National Research Council (NRC). 2008. *Research on future skill demands: Workshop summary*. Margaret Hilton, Rapporteier. Washington, DC: National Academies Press.

Nyberg, D. 1990. Power, empowerment, and educational authority. In *Educational leadership in an age of reform*, ed. S. Jacobson and J. Conway, 47–62. New York: Longman.

Olson, S. *Education Week*. 1998. Science friction. September 30.

Ophuls, W. 1977. *Ecology and the politics of scarcity*. San Francisco: W. H. Freeman.

Organisation for Economic Co-operation and Development (OECD). 2006. *Assessing scientific, reading, and mathematical literacy: A framework for PISA 2006*. Paris: OECD.

Organisation for Economic Cooperation and Development (OECD). 2007. *PISA 2006: Science competencies for tomorrow's world*. Danvers, MA: OECD.

Organisation for Economic Cooperation and Development (OECD). 2009. *Green at fifteen: How 15-year-olds perform in environmental science and geoscience in PISA 2006*. Paris: OECD.

Osborne, J. 2007. Science education for the twenty first century. *Eurasia Journal of Mathematics, Science & Technology Education* 3 (3): 173–184.

Pellegrino, J., N. Chudowsky, and R. Glaser. 2001. *Knowing what students know: The science and design of educational assessment*. Washington, DC: National Academies Press.

Peterson, N., M. Mumford, W. Borman, P. Jeanneret, and E. Fleishman. 1999. *An occupational information system for the 21st century: The development of O*NET*. Washington, DC: American Psychological Association.

Powell, J., J. Short, and N. Landes. 2002. Curriculum reform, professional development, and powerful learning. In *Learning science and the science of learning*, ed. R. Bybee, 121–136. Arlington, VA: NSTA Press.

Programme for International Student Assessment (PISA). 2003. *Problem solving for tomorrow's world: First measures of cross-curricular competencies from PISA 2003*. Paris: Organisation for Economic Cooperation and Development.

Pulakos, E. D., S. Arad, M. A. Donovan, and K. E. Plamondon. 2000. Adaptability in the workplace: Development of taxonomy of adaptive performance. *Journal of Applied Psychology* 85: 612–624.

Rakow, S. J. 1986. Teaching science as inquiry. Fastback 246. Bloomington, IN: Phi Delta Kappa Educational Foundation. ED 275 506.

Renner, J. W., M. R. Abraham, and H. H. Bernie. 1988. The necessity of each phase of the learning cycle in teaching high school physics. *Journal of Research in Science Teaching* 25 (1): 39–58.

Roberts, D. A. 2007. Scientific literacy/science literacy. In *Handbook of research on science education*, ed. S. K. Abell and N. G. Lederman, 729–780. New Jersey: Lawrence Erlbaum Associates, Inc.

Rodriguez, I., and L. J. Bethel. 1983. An inquiry approach to science and language teaching. *Journal of Research in Science Teaching* 20 (4): 291–296.

Rudolph, J. 2005. Epistemology for the masses: The origins of "the Scientific Method" in American schools. *History of Education Quarterly* 45 (3): 341–376.

Rudolph, J. 2008. The legacy of inquiry and the Biological Sciences Curriculum Study. *BSCS: Measuring our success.* Ed. R. Bybee. Colorado Springs, CO: BSCS.

Russell, G. 1998. Elements and implications of a hypertext pedagogy. *Computers and Education* 31 (2): 185–193.

Rutherford, F. J. 1964. The role of inquiry in science teaching. *Journal of Research in Science Teaching* 2: 80–84.

Rutherford, F. J. 2000. Coherence in high school science. *Making sense of integrated science: A guide for high schools.* Colorado Springs, CO: BSCS.

Schifter, D., ed. 1996. *What's happening in math class?* New York: Teachers College Press.

Schmidt, W. H., and C. C. McKnight. 1998. What can we really learn from TIMSS? *Science* 282: 1830–1831.

Schmidt, W., C. McKnight, L. Cogan, P. Jakwerth, and R. Houang. 1999. *Facing the consequences: Using TIMSS for a closer look at U.S. mathematics and science.* Dordrecht, The Netherlands: Kluwer Academic Publishers.

Schmidt, W., C. McKnight, C., R. Houang, H. Wang, D. Wiley, L. Cogan, and R. Wolfe. 2001. *Why schools matter: A cross-national comparison of curriculum and learning.* San Francisco: Jossey-Bass.

Schreiner, C., and S. Sjoberg. 2004. Sowing the seeds of ROSE. Background, rationale, questionnaire development and data collection for ROSE (The Relevance of Science Education)—a comparative study of students' views of science and science education. Dept. of Teacher Education and School Development. Oslo: University of Oslo, Norway.

Schwab, J. J. 1958. The teaching of science as inquiry. *Bulletin of the Atomic Scientists* 14: 374–379.

Schwab, J. J. 1960. Enquiry, the science teacher, and the educator. *The Science Teacher* 27: 6–11.

Schwab, J. J. 1963. *Biology teacher's handbook*. New York: John Wiley and Sons.

Schwab, J. J. 1966. *The teaching of science*. Cambridge, MA: Harvard University Press.

Secretary's Commission on Achieving Necessary Skills (SCANS). 1991. What work requires of schools: A SCANS report for America 2000, U.S. Department of Labor, xvii-xviii.

Seidel, T., M. Prenzel, J. Wittwer, and K. Schwindt. 2008. Instruction in science. Special report on PISA 2006, prepared for Australian Council for Education Research. Camberwell, Australia.

Shymansky, J. 1984. BSCS programs: Just how effective were they? *The American Biology Teacher* 40 (1): 54–57.

Shymansky, J. A., L. V. Hedges, and G. Woodworth. 1990. A reassessment of the effects of inquiry-based science curricula of the 60s on student performance. *Journal of Research in Science Teaching* 27 (2): 127–144.

Shymansky, J. A., W. C. Kyle, and J. M. Alport. 1983. The effects of new science curricula on student performance. *Journal of Research in Science Teaching* 20: 387–404.

Singer, S. R., M. L. Hilton, and H. A. Schweingruber. 2006. *America's lab report: Investigation in high school science*. Washington, DC: National Academies Press.

Spector, B. 1989. *Empowering teachers: Survival and development*. Dubuque, IA: Kendall/ Hunt Publishing.

Stedman, C. H. 1987. Fortuitous strategies on inquiry in the good ole days. *Science Education* 71 (5): 657–665.

Stillman, C. 1997. Science in an ecology of achievement. The First Paul F-Brandwein Symposium, Dingmans Ferry, PA.

Thier, H. D. 1971. Laboratory science for visually handicapped elementary school children. *New Outlook for the Blind* 65: 190–194.

Thier, H. D., and B. Daviss. 2001. *Developing inquiry-based science materials: A guide for educators*. New York: Teachers College Press.

Tikka, P., M. Kuitunen, and S. Tynys. 2000. Effects of educational background on students' attitudes, activity levels, and knowledge concerning the environment. *Journal of Environmental Education* 31 (3): 12–19.

Welch, W. S., L. E. Klopfer, G. S. Aikenhead, and J. T. Robinson. 1981. The role of inquiry in science education: Analysis and recommendations. *Science Education* 65 (1): 33–50.

Wiggins, G., and J. McTighe. 2005. *Understanding by design*. Exp. 2nd ed. Alexandria, VA: Association for Supervision and Curriculum Development.

Wilson, E. O. 2003. *The future of life*. New York: Vintage Books.

About the Author

Rodger W. Bybee is past executive director of the Biological Sciences Curriculum Study (BSCS), a nonprofit organization that develops curriculum materials, provides professional development, and conducts research and evaluation for the science education community.

Prior to joining BSCS, he was executive director of the National Research Council's Center for Science, Mathematics, and Engineering Education (CSMEE) in Washington, DC. Between 1986 and 1995, he was associate director of BSCS. He participated in the development of the National Science Education Standards, and from 1993 through 1995, he chaired the content working group of that National Research Council project. At BSCS, he was principal investigator for four new National Science Foundation (NSF) programs: an elementary school program titled *Science for Life and Living: Integrating Science, Technology, and Health*, a middle school program titled *Middle School Science & Technology*, a high school biology program titled *Biological Science: A Human Approach*, and a college program titled *Biological Perspectives*. His work at BSCS also included serving as principal investigator for programs to develop curriculum frameworks for teaching about the history and nature of science and technology for biology education at high schools, community colleges, and four-year colleges, and curriculum reform based on national standards. Dr. Bybee currently participates in the Programme for International Student Assessment (PISA) of the Organisation for Economic Co-operation and Development (OECD).

From 1990 to 1992, Dr. Bybee chaired the curriculum and instruction study panel for the National Center for Improving Science Education (NCISE). From 1972 to 1985, he was professor of education at Carleton College in Northfield,

Minnesota. He has been active in education for more than 30 years, having taught science at the elementary, junior and senior high school, and college levels.

Dr. Bybee has written widely, publishing in both education and psychology. He is coauthor of a leading textbook titled *Teaching Secondary School Science: Strategies for Developing Scientific Literacy.* His most recent books include *Achieving Scientific Literacy: From Purposes to Practices* (1997) and *Learning Science and the Science of Learning* (NSTA, 2002), and he co-wrote *Teaching Secondary School Science: Strategies for Developing Scientific Literacy* for the past eight editions.

Over the years, he has received awards as a leader of American education and an outstanding educator in America, and in 1979 he was Outstanding Science Educator of the Year. In 1989, he was recognized as one of the 100 outstanding alumni in the history of the University of Northern Colorado. Dr. Bybee's biography has been included in the Golden Anniversary 50th Edition of *Who's Who in America.* In April 1998, NSTA presented Dr. Bybee with the NSTA's Distinguished Service to Science Education Award. In 2007, he received the Robert H. Carleton Award, NSTA's highest honor, for national leadership in science education.

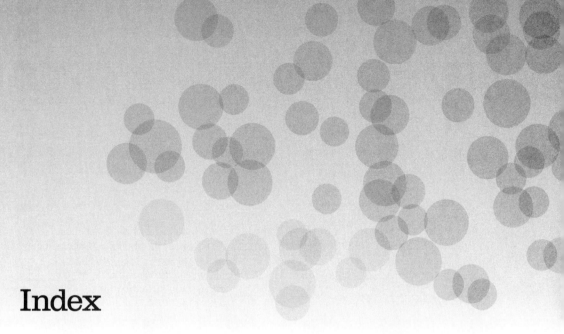

Index

Index